Fluid Inclusion Effect in Flotation of Sulfide Minerals

Fluid Inclusion Effect in Flotation of Sulfide Minerals

Shuming Wen
Professor
Department of Mineral Process Engineering
Kunming University of Science and Technology (KUST), China

Jian Liu
Associate Professor
Mineral Processing at Kunming University of Science and Technology (KUST), China

Jiushuai Deng
Professor
School of Chemical & Environmental Engineering
China University of Mining & Technology
Beijing, China

ELSEVIER

Elsevier
Radarweg 29, PO Box 211, 1000 AE Amsterdam, Netherlands
The Boulevard, Langford Lane, Kidlington, Oxford OX5 1GB, United Kingdom
50 Hampshire Street, 5th Floor, Cambridge, MA 02139, United States

Notices
Knowledge and best practice in this field are constantly changing. As new research and experience broaden our understanding, changes in research methods, professional practices, or medical treatment may become necessary.

Practitioners and researchers must always rely on their own experience and knowledge in evaluating and using any information, methods, compounds, or experiments described herein. In using such information or methods they should be mindful of their own safety and the safety of others, including parties for whom they have a professional responsibility.

To the fullest extent of the law, neither the Publisher nor the authors, contributors, or editors, assume any liability for any injury and/or damage to persons or property as a matter of products liability, negligence or otherwise, or from any use or operation of any methods, products, instructions, or ideas contained in the material herein.

Library of Congress Cataloging-in-Publication Data
A catalog record for this book is available from the Library of Congress

British Library Cataloguing-in-Publication Data
A catalogue record for this book is available from the British Library

ISBN: 978-0-12-819845-2

Publisher: Candice Janco
Acquisition Editor: Glyn Jones
Editorial Project Manager: Naomi Robertson
Production Project Manager: Vignesh Tamilselvvan
Cover Designer: Victoria Pearson

Typeset by TNQ Technologies

Contents

1. Mineral fluid inclusions 1

1.1 Definition of mineral fluid inclusions 1
1.2 Formation and mechanism of inclusions 3
1.2.1 Growth and defects of mineral crystals 3
1.2.2 Fluid capture by fluid inclusion 5
1.3 Changes in fluid inclusion after fluid capture 8
1.3.1 Phase changes 8
1.3.2 Physical changes 11
1.4 Inevitability and universality of fluid inclusions 15
Further reading 15

2. Classification of fluid inclusions 17

2.1 Genetic classification of inclusions 19
2.1.1 Primary inclusions 19
2.1.2 Pseudo-secondary inclusions 20
2.1.3 Secondary inclusions 20
2.1.4 Metamorphosed inclusions 20
2.2 Classification of physical phase states of inclusions 21
2.2.1 Fluid inclusions 21
2.2.2 Magmatic inclusions 23
References 24

3. Methods for the detection and composition study of fluid inclusions 27

3.1 General optical microscopy research on fluid inclusions 29
3.1.1 Preparation of inclusion sheets 29
3.1.2 Identification of fluid inclusions 31
3.1.3 Morphology of fluid inclusions under the microscope 33
3.1.4 Phase state of matter in fluid inclusions 38
3.2 Modern research techniques for identifying fluid inclusions 39
3.2.1 Ultraviolet light microscopy 39
3.2.2 Infrared microscopy 39
3.2.3 Scanning electron microscopy 41
3.2.4 High-resolution X-ray computed tomography 44

3.3 Determination of inclusions' salinity 44
3.3.1 Basic principle of determining salinity of inclusions by the freezing method 45
3.3.2 Notes on the determination of salinity in inclusions by the freezing method 46
3.3.3 Instruments for measuring the freezing point 47
3.4 Extraction and analysis of fluid inclusion components 48
3.4.1 Issues to be aware of in composition analysis of inclusions 49
3.4.2 Analysis of chemical composition of group inclusions 50
3.4.3 Analysis of the chemical composition of individual inclusions 57
References 66
Further reading 67

4. Internal composition of mineral fluid inclusions 69

4.1 Gaseous-phase composition of inclusions 70
4.2 Liquid-phase components of inclusions 74
4.3 Solid-phase composition of inclusions 78
4.4 Metal components in fluid inclusions 80
4.4.1 Analysis and calculation of heavy metal elements in ore-forming fluids 80
4.4.2 Metal components in fluid inclusions in Cu–Au mineral 83
4.4.3 Metal components in fluid inclusions of Pb–Zn mineral 85
4.4.4 Metal components of group inclusions in quartz and pyrite 88
References 95
Further reading 96

5. Component release of fluid inclusions in sulfide mineral 97

5.1 Analysis of mineral raw materials 99
5.1.1 Chalcopyrite and associated minerals 99
5.1.2 Sphalerite and quartz 100
5.1.3 Galena 102
5.1.4 Pyrite 102
5.2 Research methods for fluid inclusions in sulfide mineral 105
5.3 Morphology and component release of fluid inclusions in chalcopyrite 107
5.3.1 Infrared optical microscopic analysis of the fluid inclusions in chalcopyrite 107
5.3.2 SEM/EDS of the position of fluid inclusions in chalcopyrite 108
5.3.3 HRXMT analysis 111
5.3.4 Component release of the fluid inclusions in chalcopyrite 111
5.4 Component release of the fluid inclusions in associated minerals of chalcopyrite 114
5.4.1 Component release of the fluid inclusions in the associated quartz and calcite of chalcopyrite 114
5.4.2 Component release of fluid inclusions in chalcopyrite, sphalerite, and associated minerals 118

5.5 Morphology and component release of fluid inclusions in sphalerite and associated quartz 123
5.5.1 Morphology and types of fluid inclusions in sphalerite and associated quartz 123
5.5.2 Freezing point and salinity value of the fluid inclusions in sphalerite 126
5.5.3 SEM/EDS analysis of the fluid inclusions on sphalerite surface 127
5.5.4 Component release of fluid inclusions in sphalerite and quartz 129
5.6 Morphology and component release of fluid inclusions in galena 133
5.6.1 Infrared optics and SEM/EDS of fluid inclusions in galena 133
5.6.2 Component release of inclusions in galena 134
5.7 Morphology and component release of fluid inclusions in pyrite 134
5.7.1 Infrared optical microscopic analysis of fluid inclusions of pyrite in Weixin 134
5.7.2 SEM/EDS detection of the positions of inclusions in pyrite from Weixin 137
5.7.3 High-resolution X-ray topography of microfault pyrite from Weixin 140
5.7.4 Component release of the fluid inclusions of pyrite from Weixin 141
5.7.5 Component release of quartz fluid inclusion in the pyrite deposits of Weixin 143
5.7.6 Component release of fluid inclusions of pyrite from the polymetallic sulfide deposits of Dapingzhang 145
5.7.7 Component release of the fluid inclusions of quartz from polymetallic sulfide deposits of Dapingzhang 147
Further reading 149

6. Solubility of sulfide mineral and chemical behaviors of solution after release of inclusion components 151

6.1 Solubility of sulfide mineral 152
6.1.1 Research methods 152
6.1.2 Solubility of the chalcopyrite surface 153
6.1.3 Solubility of the sphalerite surface 156
6.1.4 Solubility of the pyrite surface 158
6.2 Equilibrium theory calculation of solubility of sulfide minerals 160
6.2.1 Solubility of sulfide mineral in pure water 160
6.2.2 Solubility of sulfide mineral at different pH 164
6.3 Chemical equilibrium calculation of metal ions in slurry solution 169
6.3.1 Cu^{2+} taring and component distribution 169
6.3.2 Zn^{2+} taring and component distribution 172
6.3.3 Pb^{2+} taring and component distribution 173
References 177

7. Interactions among components of fluid inclusions in sulfide mineral, mineral surfaces, and collectors 179

7.1 ζ potential measurement of the adsorption of released components of inclusions on mineral surfaces 180

7.1.1 Test method for ζ potential 181

7.1.2 Surface adsorption of inclusion components in chalcopyrite 182

7.1.3 Surface adsorption of inclusion components in sphalerite 186

7.1.4 Surface adsorption of inclusion components in pyrite and galena 188

7.2 Density functional theory of the interactions between the components of fluid inclusion colonies and mineral surfaces 190

7.2.1 Crystal model and calculation method 190

7.2.2 Surface relaxation and reconstitution of sulfide mineral 195

7.2.3 Interaction of chalcopyrite surface and the copper in inclusions 209

7.2.4 Interaction of sphalerite surface and the copper in inclusions 211

7.2.5 Interaction of the pyrite surface and the Cu in inclusions 218

7.3 Interactions among sulfide mineral surface, components of fluid inclusions, and collectors 219

7.3.1 Interaction of chalcopyrite surface, Cu components, and xanthate 220

7.3.2 Interaction of sphalerite surface, Cu, and xanthate 223

7.4 Interaction of the components released from mineral fluid inclusions in sulfide mineral flotation 226

Reference 229

Further reading 229

Index 231

Chapter 1

Mineral fluid inclusions

Chapter outline

1.1 Definition of mineral fluid inclusions 1
1.2 Formation and mechanism of inclusions 3
1.2.1 Growth and defects of mineral crystals 3
1.2.2 Fluid capture by fluid inclusion 5
1.2.2.1 Irregular crystal growth 5
1.2.2.2 Unbalanced growth of various parts inside the crystal 6
1.2.2.3 Inclusion formation through differences in the medium concentration 6
1.2.2.4 Discontinuity of crystal growth 6
1.2.2.5 Effects of solid-phase substances and impurities 6
1.2.2.6 Influence of temperature and pressure 7
1.3 Changes in fluid inclusion after fluid capture 8
1.3.1 Phase changes 8
1.3.1.1 Crystallization on the body wall 8
1.3.1.2 Daughter minerals 10
1.3.1.3 Shrinkage and immiscibility 10
1.3.1.4 Metastability 11
1.3.2 Physical changes 11
1.3.2.1 Volume and shape changes 12
1.3.2.2 Fluid infiltration and loss 13
1.4 Inevitability and universality of fluid inclusions 15
Further reading 15

1.1 Definition of mineral fluid inclusions

The research into mineral fluid inclusions originated in geology in the mid-19th century. Early observations of inclusions were made in minerals such as quartz and topaz, and this discovery preluded the study of inclusions in minerals. After in-depth research, a more comprehensive and scientific definition of fluid inclusions has been put forward.

"Mineral fluid inclusions" refers to the diagenetic and ore-forming solutions that are captured in mineral crystal defects, holes, lattice vacancies, dislocations, and microcracks during the growth/formation of mineral crystals.

Fluid Inclusion Effect in Flotation of Sulfide Minerals. https://doi.org/10.1016/B978-0-12-819845-2.00001-6

The fluid is sealed in the host crystal, as an independent closed system of the phase boundary. To understand this definition, the following geological terms are carefully expanded:

(1) The diagenetic and ore-forming fluids in the definition refer to the fluid media, such as magma, solution, and gases around the main minerals, from which the inclusions are captured. They do not include the media's debris such as crystal chips, cuttings, and crystals.
(2) Host crystals contain the inclusions, and are formed simultaneous with the inclusions.
(3) When the fluid captured in the inclusions is a supersaturated solution, a solid phase of the daughter mineral can be crystallized from the solution when the temperature decreases and be enclosed in the inclusion (see Fig. 1.1). The inclusions containing daughter minerals (sealed within the inclusion) coexist with bubbles and liquids as well.
(4) Regarding the phase boundary between the inclusion and the host crystal, the outer contour of the inclusion observed under the microscope is the phase boundary between the inclusion and the host crystal. During the ore formation, due to changes in pressure, temperature, etc., the host crystal produces various crystal defects such as caves and cracks that capture fluids, which are later sealed within the cracks. Therefore, the outer contour of the inclusion is the equilibrium boundary interlayer between the inclusion and the host crystal.
(5) After the inclusions are captured during the growth of the host crystals, no further material exchange occurs with the outside; thus the phase boundary with the host crystal becomes an independent system. After the inclusions are captured during the growth of the host crystals, no further material exchange occurs with the outside. In addition, the inclusions have a phase boundary with the host crystal and thus become an independent

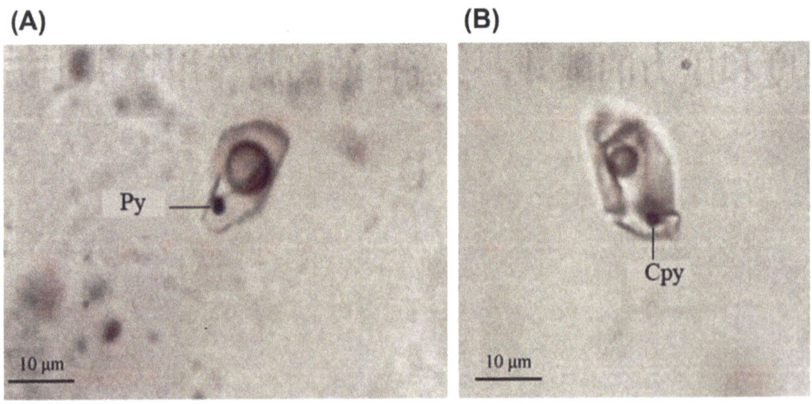

FIGURE 1.1 Fluid inclusions rich in daughter minerals of quartz veins from the Badi copper ore deposits in Shangri-La, Yunnan, China. (A) Inclusions containing pyrite (*Py*) crystals. (B) Inclusions containing chalcopyrite (*Cpy*) crystals. *According to Su J, Geochemical characteristics of fluids of the Shangri-La Badi copper deposits. 2014.*

system. The inclusions exist together with the host crystals and are reserved to now.

Geological studies have shown that the sizes of fluid inclusions are closely related to the prevailing geological conditions and processes during their formation, and are different in different minerals and in the same minerals. The length of inclusions is usually around 10 μm, and most are less than 100 μm. According to the definition of inclusions, mineral fluid inclusions can be regarded as an independent geochemical system containing samples of diagenetic, ore-forming fluids of a specific geological period. The system has the following characteristics: First, it is enclosed—after the capture of inclusions in the host, no substance enters or escapes. Second, the system is homogeneous—the substances are captured in the homogeneous phase when the inclusions are formed; however, studies have shown that there are inclusions captured with heterogeneous systems. Third, the system is an isovolumetric system—the volume does not change after the inclusions are formed.

1.2 Formation and mechanism of inclusions

1.2.1 Growth and defects of mineral crystals

Generally, there is no perfect naturally formed crystal in the world, and certainly none has been obtained even under the most stringent experimental conditions. Any condition interfering with the growth of intact crystals can result in crystal defects, which causes the formation of inclusions. In any kind of fluid medium, irregular crystal growth or recrystallization often leads to crystal defects. A small amount of fluid medium, after penetrating these defects, is sealed, forming the fluid inclusion. The mechanism by which fluid inclusions are formed is a process that goes through defect formation, fluid penetration, and crystal regrowth and closure during crystal growth or recrystallization.

The formation of mineral solid crystals is a transformation from the gas phase, the liquid phase, and the silicate molten mass phase to the solid phase. The particles are stacked in a three-dimensional space according to a specific law, and the essence is the process from irregular to regular arrangement of the particles. Ideally, when the ore-forming solution or molten mass of magmatic silicate reaches supersaturation due to the temperature decrease, the particles of the medium are polymerized into microcrystal nuclei or grains of crystallization according to certain rules to form a crystal center that eventually grows into the crystal. Generally, when a nucleus is formed, the particles adhere to the crystal nucleus in accordance with the predetermined crystal structure/template for continuous growth.

Taking the ideal crystal growth process in Fig. 1.2 as an example, the three-sided concave position 1 is the most attractive to the particles. Thus, particles are stacked to this position first, maximally releasing the crystal energy to minimize the internal energy of the crystal and maintain a stable state.

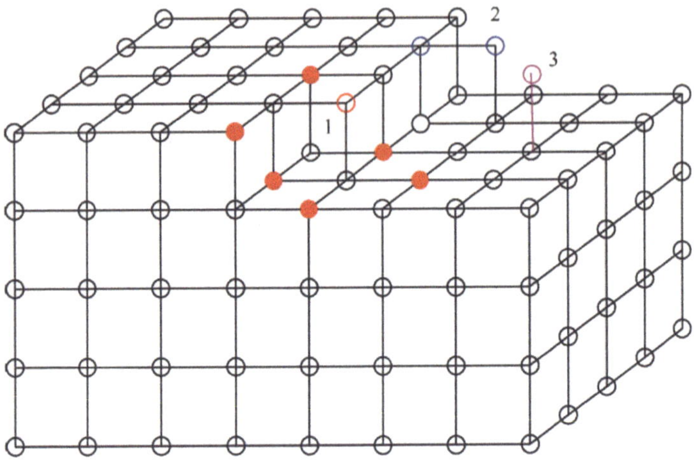

FIGURE 1.2 Sequence of particle stacking during ideal crystal growth. Position 1 is the three-sided concavity; position 2 the two-sided concavity; position 3 is the general position. *According to Pan Z, Crystallography and mineralogy. 1985.*

Particles are also stacked in the two-sided concave position 2 in Fig. 1.2, and finally to the position without concavity (position 3 in Fig. 1.2). In the ideal growth process of a crystal, a layer of mesh, a bedrock of crystal nucleus, is formed. Thereafter, the adjacent layer starts growing (through the accumulation of the crystal particles) and gradually shifts outward. A common row of two adjacent meshes forms the crystal edge; the entire crystal is surrounded by a crystal face, eventually forming a crystalline polyhedron.

Understanding the genesis and mechanism of inclusion formation is essential to explain the chemical composition, environmental conditions (pressure, volume, temperature, etc.), and other information about inclusions. In general, crystal growth is controlled by the physical and chemical properties of the diagenetic ore-forming solution, the complex interactions between them, and the environmental conditions (external pressure, contraction and expansion of fluid, and temperature); after the growth of the crystal, it is inevitably affected as geological action evolves. During the geological process, crystal growth is largely affected by the environment. Growth that deviates from the ideal conditions produces defects in the localized range of the crystal structure. According to the extension of the defect in the space, it can be divided into the following cases:

(a) point defects: lattice impurity atoms, vacancies, void atoms, etc.
(b) line defects: dislocations
(c) volume defects: cracks, voids, etc.

The formation of inclusions is closely related to the defects mentioned in the crystal growth process, especially the volume defects. The formation of volume defects constitutes the main space where diagenetic, ore-forming fluids are captured in mineral crystals. Usually, a considerable number of inclusions formed are due to volume defects, because the point and the line defects can dissolve, expand, and erode during the crystal growth, giving rise to the volume defects.

1.2.2 Fluid capture by fluid inclusion

Any factor that hinders crystal growth during mineral formation can cause crystal defects, which create a prerequisite for capturing fluids in crystals. After capturing the fluids, the defects continue to grow to form inclusions, as summarized in the following.

1.2.2.1 Irregular crystal growth

During the development of mineral crystals, a large number of irregular structures (such as mosaic structure, random orientation of crystals in the cluster, and screw dislocations) including a variety of voids are formed, which result in inclusions. Microscopic studies have shown that inclusions are easily formed by fluids between two large adjacent spirals or at the center of the spirals (see Fig. 1.3B). The crystals, formed from defects such as cracks in this growth period, often lead to irregular growth. It is easy to form new defects on the basis of the original cracks, which then capture fluids to form inclusions (see Fig. 1.3C).

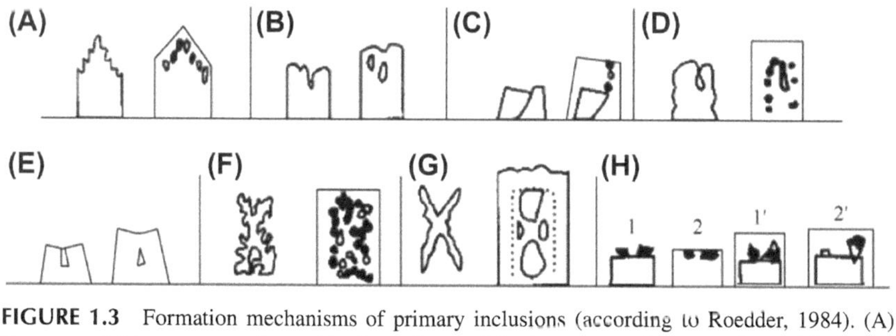

FIGURE 1.3 Formation mechanisms of primary inclusions (according to Roedder, 1984). (A) Dense crystal layer covers the rapid growth layer of the branches and forms a layered inclusion colony. (B) Inclusions captured between growth spirals or at the center of the growth spirals. (C) Crystal face cracks, resulting in poor growth of crystals that tend to form inclusions. (D) Parts of the crystal melt, creating pits and curved crystal faces. Inclusions are captured through crystal regrowth. (E) The subparallel growth of the structural units of a crystal, with the captured inclusions. (F) Due to the decrease in temperature, the magma becomes supersaturated to a certain phase but fails to nucleate. When the nucleus finally appears, it grows rapidly, forming skeletons or dendrites. (G) When the supersaturation is reduced it forms a dense layer, which becomes surrounded to form an inclusion. (H) Crystal corners and edges grow rapidly, forming pits that can capture large inclusions with solid debris falling on the growing crystal face. The solid debris is either wrapped or pushed toward the growth frontier so that inclusions 1 and 1′ are formed when the solid particles are buried by the growing crystal faces.

1.2.2.2 Unbalanced growth of various parts inside the crystal

Crystal growth is accomplished by the continuous supply of fluid media to the growth face. There are two ways: replenishment by fluid diffusion and mass flow through the fluids. It is undeniable that replenishment will have certain concentration differences in the different parts of the cross sections of the crystal, that is, a concentration gradient will exist. Obviously, the edges and corners, as well as the angular points of the crystal, are easily replenished by the solutes in the fluids, while the center of the crystal, which has little contact with the fluids, is less prone to replenishment. Wilkins and Barkas (1978) confirmed that under certain conditions, the crystals formed from the same supersaturated solution are different. The smallest crystal face center and the largest angular point create the difference in the growth rates of different parts of the crystal. Large pits in the crystal edge and the angular points facilitate the capture of ore-forming solutions or diagenetic molten mass to form inclusions (See Fig. 1.3G).

1.2.2.3 Inclusion formation through differences in the medium concentration

The mother liquor has a certain pressure, temperature, and concentration, and physical and chemical parameters in the growth constantly change and restrict one another. Both the physical and the chemical parameters are mutually constrained and constantly change during growth. The concentration of the growth medium is not constant; for example, the decrease in temperature causes a change in the concentration of the medium as well as an increase in the supersaturation of the solution. Generally, a highly supersaturated solution will accelerate the growth of the crystal and cause dendritic growth with a concave angle. In addition, studies have shown that the concentrations of certain minor components in the solution can have a great impact on the perfection of the crystal growth.

1.2.2.4 Discontinuity of crystal growth

Crystal grows in stages, or even in many stages, of intermittent growth within certain periods. Temporary interruptions and changes in the principal growth conditions can lead to crystal defects. For example, due to the intermittent regrowth of a crystal, the early crystal faces formed are easily dissolved and etched to form pits. Some deeper concave pits form fluid inclusions along the crystal face when the fluids are refilled (see Fig. 1.3D).

1.2.2.5 Effects of solid-phase substances and impurities

One of the causes of crystal defects is that solid matter or impurities adhere to crystals that are crystallizing (Fig. 1.3H). The attachment of solid materials takes up the original growth space and hinders the recharge of the ore-forming solution; the ore-forming fluid has to flow around the solids in a streamlined motion, which causes the unbalanced flow and slow flow rate of the solutions.

The defects are formed to capture the ore-forming solutions, thus giving rise to inclusions.

1.2.2.6 Influence of temperature and pressure

During mineralization, changes in physical conditions such as temperature and pressure are major important causes of crystal defects and voids. Changes in temperature and pressure could induce a series of interlocking changes such as solution concentration, solubility, saturation, and crystallization rate. Undoubtedly, these changes inevitably lead to the irregular growth of the crystals and the formation of defects, which cause a large number of inclusions (Fig. 1.3F).

Based on this analysis, the main mechanisms of inclusions in minerals can be grouped into four categories: (1) changes in the crystal growth mechanism; (2) changes in the concentrations of certain components in the solution; (3) changes in the crystal growth rate; (6) interaction between the solid-, liquid-, or gaseous-phase particles and the crystal face growth. The different mechanisms of formation result in different types of inclusions. Whereas mechanisms (1), (2), and (3) result in the formation of a fluid inclusion group, thus exhibiting growth bands in the crystal, mechanism (4) often leads to the formation of isolated inclusions, and the captured impurities are visible under the microscope in some isolated inclusions.

It is generally believed that inclusions are formed by the capture of uniform fluids, and the fluid in the inclusions represents the geochemical system during the mineralization. However, research data from the study of inclusions over the years have shown that a considerable proportion of inclusions is not formed from homogeneous fluid systems. In other words, inclusions are also formed with heterogeneous fluid phases. These inclusions constitute:

(1) inclusions captured from a liquid + gas system. The heterogeneous fluid system of liquid + insoluble gases can be generated by changes in environmental influences, such as temperature, pressure, and expansivity. The original uniform fluid becomes immiscible due to temperature and pressure differences; for example, the fluid boils due to pressure release or the temperature rises, causing the release of gases. Studies show that inclusions in some stalactites of limestone caves are formed by capturing fluids from nonuniform or immiscible systems. During the formation of stalactites, liquids and gases are present in the environment, and the inclusions are formed in this mixed system. When the inclusions are formed, the gases and liquids are simultaneously captured and enclosed.

(2) inclusions captured from a liquid + solid system. Usually, the region of mineral growth is filled with fluids, crystals, or some tiny solid materials. When these solid-rich fluids are sealed in the mineral, liquid + solid heterogeneous-phase inclusions are formed. The wrapped crystals or solid fragments are called "daughter minerals," and such inclusions are said to contain "daughter minerals."

(3) inclusions captured from two immiscible liquids ($L_1 + L_2$). The two immiscible fluids may be completely or partially immiscible, they include systems such as oil and water, water and CO_2, and molten mass and fluids. When studying the Mississippi Valley–type Pb–Zn deposits, researchers found that the fluorites contain fluid inclusions of oil and water. This indicates that the fluids enclosed in the Pb–Zn mineral inclusions mixed with the oil and water of oil fields when passing through the Mississippi basin before forming the Pb–Zn deposits. The oily inclusions are a good indicator of oil in the search for it.

Another example of $L_1 + L_2$ inclusions is the hydrothermal fluid inclusion system. Here the hydrothermal fluid is separated from the magma, and coexists with the molten mass. The inclusions captured at this stage contain two immiscible phases, the fluid phase and the molten mass. Huanzhang Lu discovered the molten fluid phase in his study of inclusions in granites obtained from Tibet. This type of inclusion provides evidence for the separation of fluids from magma, which is an important evolution of magma, and the nature of hydrothermal fluids, etc.

1.3 Changes in fluid inclusion after fluid capture

Most fluid inclusions undergo certain changes after fluid capture. Their shapes and physical and chemical properties are significantly different from those before the capture, and this phase change is noticeable. Homogeneous-phase inclusions captured at high temperatures inevitably undergo phase changes when cooled. This phase change can be restored by warming or cooling in the lab. That is, the process is reversible, thus providing valuable data on the temperature, pressure, and composition of the inclusions. On the other hand, the physical changes after fluid capture are difficult to recover. Physical changes in inclusions play an important role in geological studies, unlike in mineral processing, because they greatly enhance the interpretation of PTX (pressure, temperature, and fluid composition) data obtained from phase studies. The phase and physical changes in fluids after capture are specifically described as follows.

1.3.1 Phase changes

The small amount of fluid in mineral crystals (inclusions) can undergo various phase changes and physical changes also. The originally sealed single-phase homogeneous fluid can become a multiphase inclusion at room temperature. The newly formed phase in the inclusion is called the subphase, and when it crystallizes out of solution, it is called the daughter crystal or daughter mineral.

1.3.1.1 Crystallization on the body wall

The solubility of solid materials in nature increases with increasing temperature. Therefore, when the fluid inclusions captured under high temperature are naturally cooled, crystals of the host minerals are inevitably formed. This is because the fluid contains a saturating amount of the host mineral components. Usually, the crystallization occurs within the inclusions and does not form separate host crystals in the inclusions. Also in fluid inclusions, a small amount of other primary minerals typically crystallizes on the body walls, since most minerals have low solubility at moderate to low temperatures. For example, in quartz, SiO_2 has a very low solubility, so it is easily crystallized out of solution. Most fluid inclusions do not dissolve from the body walls of the inclusions when heated, and are not crystallized when the temperature is lowered. However, at a higher temperature (>500°C), there is accidental dissolution. This implies that in fluid inclusions, there is generally no crystallization on the body walls of the inclusions. There is no increase in the volume of the body cavity caused by the dissolution of the body wall due to rising temperature. However, for soluble minerals such as halite, the solubility increases sharply during heating, and the body wall of the inclusion dissolves due to the rising volume of the body cavity.

When molten silicate inclusions are warmed, their body walls are dissolved. This is because as they cool, the molten silicic acid in the inclusions crystallizes a considerable amount of the host crystals deposited on the body wall, forming a thick lining. Roedder (1984) explained this phenomenon in his book (see Fig. 1.4), and described the traces formed by the lining of the body wall of the molten silicate inclusions after capturing the inclusions. When the

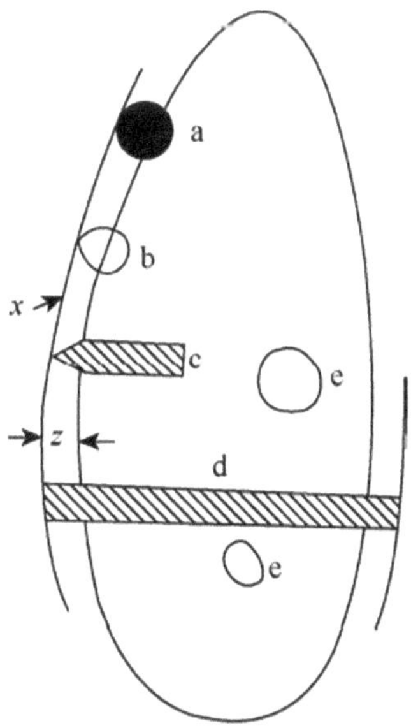

FIGURE 1.4 Crystallization of the host crystal layer (lining) on the body wall of a molten silicate inclusion after capturing fluid. *According to Roedder, Fluid inclusions [M]. Mineralogical society of America. 1984.*

inclusion is captured, the body wall is assumed to be at *x*, and an immiscible sulfide spherule (a), a bubble (b), and two subminerals (c and d) are nucleated on the wall. The thickness of the lining *z*, and the daughter crystal c that crystallized before the lining formed, is affected by the crystallization of the host crystal layer on the body wall. The daughter crystal d grows after the nucleation, and it is not affected by the crystallization of the host crystal layer on the body wall. After it grows into a platelike crystal, the host crystal is precipitated on the body wall. The boundary *x* is usually invisible, so it is difficult to determine if there are no subphases enclosed within the lining. However, it is still possible to determine the inclusions in the lining even if the evidence shows that the silicate inclusions have no lining, as in the lack of a subphase buried in the body wall, the crystallization of the host crystal layer on the body wall of inclusions is still certain. This information is important for studying the composition of molten silicate inclusions.

In addition, in the inclusions of basalt olivine, there is nucleation of the immiscible sulfide spherules or the daughter crystals on the body wall, with the host crystal growing around them. The bubbles may differ from the sulfide spherules even if nucleating on the wall. The olivine grows under the bubbles, and pushes them into the residual melt. Therefore, the bubble (e) is rarely wrapped on the body wall.

1.3.1.2 Daughter minerals

The original uniform fluids in inclusions may contain a saturated amount of the host crystals or other mineral components during it capture. When a component reaches saturation, crystals called daughter crystals or daughter minerals are precipitated as the temperature reduces. When the inclusions are warmed in the laboratory, the soluble crystals redissolve in the fluid, and the temperature at which the daughter crystals are completely dissolved is the solution temperature. The most common daughter crystals are cubic stone halite (NaCl), followed by sylvine, carbonates, oxides, silicates, sulfates, and sulfides. Most fluid inclusions with soluble daughter minerals contain only monocrystals of the subphases. If the daughter minerals in the fluid inclusions are polycrystalline, their solubility in the liquid phase is very low. For example, the large inclusions in Brazilian pegmatite contain several small pyrite cubes, and the solubility of the pyrite in the liquid phase is very low.

1.3.1.3 Shrinkage and immiscibility

The uniform-phase fluids in inclusions usually have a much higher shrinkage rate than the host crystals during cooling. Most minerals have a coefficient of thermal expansion that is 1–3 orders of magnitude smaller than the coefficient of thermal expansion of water. When cooled to room temperature, the volatiles in the inclusions concentrate to form bubbles. The internal pressure of such a bubble varies depending on the fluids in the inclusions. The vapor pressure of the bubble in nonvolatile fluids is low; in the nonvolatile molten silicate inclusion the internal pressure is close to vacuum. Inclusions rich in volatile substances (CO_2, CH_4, etc.) are easily concentrated into bubbles due to volatilization. The bubbles may exist as thick fluids, and if the density of the bubble is greater than that of the liquid, the "bubble" sinks into the liquid. If sufficiently dense, the separated CO_2-rich fluids are separated into two fluids, liquefied and gaseous CO_2. The liquefaction occurs below the critical temperature of pure CO_2 (31.1°C). If the fluid density deviates from the critical density of CO_2 or other gases (N_2 and CH_4), the critical temperature of CO_2 can be lowered.

During cooling, if the melt in the molten silicate inclusion is iron-saturated sulfide, a sulfide spherule can be formed. The resulting inclusion will contain immiscible gases and immiscible sulfide melt, which enriches the nickel and steel from silicate melt. This discovery provides clues to the formation of magmatic sulfide deposits. During the cooling, the molten silicate inclusions containing volatile matter may form several fluid phases due to the immiscibility, such as glass, saline solution, liquid CO_2, or other gases. They are arranged in the order of minimum surface energy in inclusions, and the order from the edge to the center is as follows: glass, saline solution, liquid CO_2, and gases. This can be used as an auxiliary sign to identify them.

1.3.1.4 Metastability

In a group of inclusions with the same origin and different sizes, the phases may be different. The phases of small inclusions are simple, while those of large inclusions are often complicated. The different phases are not caused by the captured heterogeneous-phase media. The fluids of small inclusions do not easily form stable nuclei, and even small single inclusions certainly do not form stable bubbles under negative pressure after hundreds of years. They are metastable systems due to the smallness of the inclusions and a few nucleations. In large systems, there are more inclusions and more chances of nucleation; hence they are stable systems. The image difference exhibited in the metastable systems is not a result of capturing heterogeneous phases during the formation of the inclusion.

1.3.2 Physical changes

A series of physical changes occurs after fluids are captured in inclusions, including changes in volume, shape, position, and fluid infiltration and loss. After formation, the irregular inclusions gradually transform into regular inclusions due to their tendency to minimize surface area. The flat, long plate-shaped inclusions become transformed to oval or round. Under the influence of thermal gradient and stress gradient, the substances on one side of the inclusion can be dissolved, and those on the other side precipitated, thereby causing the inclusion to move.

1.3.2.1 Volume and shape changes

The volume change after the formation of the inclusions can be divided into two types:

(a) Reversible volume change: This volume change occurs in the inclusion under the influence of external conditions. However, the volume can be restored when the external action disappears, so it is regarded as the constant or the ideal volume.

(b) Irreversible volume change: When the volume of inclusion changes irreversibly for some reason, it does not return to its original volume. In geology, reversible volume change of the inclusion has no effect on the homogenization method, and irreversible volume change cannot determine the true (original) homogeneous temperature of the inclusion. The main reasons the volume of inclusions can remain constant include the following:

  (1) After the inclusions are captured, the crystallization of the solution on the body wall or in the fluid due to cooling can be offset by heating the host crystals.

  (2) During cooling, the volume change caused by thermal contraction of the host crystals and the daughter minerals can be offset by the reversible change in thermal expansion. Ermakov, the former Soviet scholar, indicated that the daughter minerals and the crystallization on the body walls of inclusions cause a shrinkage in the body cavity of the inclusions, and therefore, the temperature measurement, which is determined by the solute in the fluids, greatly deviates. For example, the solubility of SiO_2 in solutions at low temperatures is low; thus the lining formed by cooling will be very thin in the gas–liquid inclusions of quartz. There is almost no change in the volume of the inclusions, so its effect on homogenization method is negligible. However, the situation is quite different for molten silicate inclusions. When the temperature drops, the formation of subminerals in the solution causes the volume of the inclusions to reduce. During heating, they (the lining and daughter minerals) are dissolved, and the volume of the inclusions is restored to its original size. The change in volume does not affect the results of the homogenization method. Volume changes in saline minerals such as NaCl and KCl are quickly restored since they are easily crystallized by cooling, and dissolved by heating. However, for silicate minerals, the dissolution is very slow, especially for the lining of the body wall, and since there is no clear boundary between the lining and the host crystal, it is difficult to establish judgment criteria as to when the dissolution is completed. Although the balance can be restored under a constant supply of heat, it is necessary to maintain a constant heating rate and temperature.

For the inclusions with reversible change in volume, the volume change is recoverable. The problem of the lining can be solved by controlling the experimental conditions. However, it should be noted that a permanent or irreversible change can be rendered in the volume of inclusions under one of the following conditions:

(1) "Necking down" of inclusions occurs due to recrystallization of minerals. Here the original inclusion becomes divided into several inclusions, thus changing its volume.
(2) In the case of metamorphism, the inclusions can be combined and reseparated. Some scholars have shown that gas–liquid inclusions uniformly distributed in anhydrite minerals combine into arc-shaped semicircular inclusions at 500°C, and form elliptical inclusions at 700°C. Swanenberg in 1980 also observed a similar phenomenon in middle- and high-grade metamorphic rocks in southwestern Norway, where small inclusions disappear and join together to form larger inclusions.
(3) When the ambient temperature of the inclusions rises, the inclusions may rupture due to the difference in external pressure, which causes the volume of the inclusions to increase and simultaneously release some pressure. The fluid components released from the inclusions flow into the surrounding crystal cracks, and sometimes these cracks are healed, forming satellite-like secondary small inclusions that are less dense than the original ones.
(4) Inclusions in plastic minerals may undergo permanent changes in volume due to internal or external pressure. For example, when an inclusion of halite encounters a sharp rise in temperature, a burst occurs when the temperature exceeds the homogeneous temperature of the inclusion. However, if the temperature is slowly raised, even if the homogeneous temperature is exceeded, the internal pressure will increase drastically without cracks. This is because halite has a high plasticity index; that is, when the internal pressure increases due to the rising temperature, the volume of the inclusion becomes large due to simple expansion. In addition to halite, many other minerals have plasticity, especially quartz. Volume change in inclusions induced by the plasticity of the mineral is a common occurrence. When the internal pressure exceeds the external pressure, it increases the volume of the inclusion. When the internal pressure is less than the external pressure, the volume is reduced due to the compression of the inclusion.

1.3.2.2 Fluid infiltration and loss

In general, most mineral inclusions are in closed systems, and there is no material exchange with the outside world. In other words, the components of most crystals are impermeable. Studies have shown that only hydrogen can easily pass through many minerals. Anderson and Sans studied the silicate molten mass inclusions in olivine and pyroxene from Mount Shasta in

California. It was found that when the sample is heated in a vacuum, a brown halo is formed on the body wall of the inclusion, and the moisture disappears within 18 h. This fact indicates that at higher temperatures water decomposes into hydrogen and oxygen, and hydrogen escapes by diffusion. The ferrous ions in the host crystals of olivine and pyroxene are converted to ferric ions when they react with oxygen by autoxidation.

In fact, the escape of substances in mineral inclusions occurs mostly during metamorphism. Moreover, when microcracks are produced by certain mechanisms they are taken as channels, which cause changes in the inclusions. However, natural burst results in the complete loss of inclusions. When the internal pressure of the inclusions is greater than the external pressure to a certain extent, the body walls crack. This cracking may be caused by a sudden decrease in the external pressure (as in the eruption of magma) or the external heating of a geological body. During an eruption, the inclusions may burst and discharge through a large opening, and the volatiles released flow into the magma to form empty inclusions. When the inclusions are small, it can cause only local cracking of the body walls due to decompression. The vacuum expansion after the cracking causes the internal pressure to be significantly lowered, thus preventing further cracking. The released contents of the inclusions flow into the cracks, which may heal and recrystallize to form secondary inclusions, and the resulting old and new inclusions have lower density and pressure than the original inclusions because a portion of the latter enters the cracks. There are many examples of cases in which foreign substances are added to the inclusions. For example, in plagioclase, the molten silicate inclusions are unevenly distributed. The melt solidifies to clay at the positions where the microcracks cut through the inclusions. The plagioclase of the Yangchuling granodiorite porphyry in Jiangxi exhibits this phenomenon. In fluid inclusions, when the internal pressure is lower than the external pressure, foreign substances flow into the system through the microcracks, thus changing the components of the original inclusion colony.

In geology, people study inclusions to obtain information on the chemical environment and physical and chemical conditions of diagenesis and mineralization, with strict requirements for the representativeness of the inclusion samples. This is because of the diversity and complexity of inclusions, which record the conditions of the growth and evolution of minerals. The genetic significances are different, and not all inclusions provide valid information. To select qualified samples from complex and diverse inclusions, three basic assumptions or preconditions for the study of inclusions have been outlined in geology. They are homogeneous system, closed system, and isometric system. However, in mineral processing, researchers are more concerned about the chemical composition and content of the fluids in the inclusions, as well as the surface adsorption and the chemical reaction of solutions after cracking and release of the contents of the inclusion.

1.4 Inevitability and universality of fluid inclusions

Based on the mechanism of formation and the definition of fluid inclusions, we can conclude that the existence of fluid inclusions in crystals of natural deposits is inevitable and universal. This is because there are no perfect crystals in nature; as long as there are crystal defects, fluid inclusions are inevitable. The size and quantity of fluid inclusions in minerals are closely related to the ore-forming conditions and the subsequent geological evolution. This shows that regardless of the mineral particle size or type, whether transparent or opaque, whether of natural or artificial synthesis, every crystal has fluid inclusions. There are a large number of fluid inclusions in natural minerals, and it is estimated that the number of minerals per cubic centimeter can reach 10^6-10^9, and the abundance (volume concentration) in minerals reaches 0.1% −1%. Modern scientific research has confirmed that inclusions exist in all minerals on the earth and other planets. Large inclusions, up to a few millimeters in size, can be identified by the naked eye; however, most inclusions can be clearly observed only with an appropriate optical microscope. Inclusions that can be found under a microscope are typically between a few and tens of micrometers, and inclusions less than 1 μm are also found in minerals.

Fluid inclusion is one of the most active research directions in earth science. Its research results are widely used in the study of mineral deposits, structural geology, petroleum exploration, internal fluid migration and evolution of magmatic systems, paleontology, paleoclimate environment, etc. In the mineral processing discipline, however, research on inclusions is still in its preliminary stage.

Further reading

[1] Zhang W, Chen Z. Fluid inclusion geology. Wuhan: China University of Geosciences Press Co. Ltd; 1993.

[2] Lu H, Fan H, Ni P. Fluid inclusions. Beijing: Science Press; 2004.

[3] Roedder E. Fluid inclusions. Mineralogical Society of America; 1984.

[4] Institute of Geochemistry, Chinese Academy of Sciences. Advanced geochemistry. Science Press; 1998.

[5] Xiao R, Zhang Z, Chen H, et al. Natural types of geological fluids and types of ore-forming fluids. Geosci Front 2001;8(4).

[6] Carpenter AB. Origin and chemical evolution of brines in sedimentary basins[C]//SPE Annual Fall.

[7] Roedder E. Fluid inclusion evidence for immiscibility in magmatic differentiation. Geochem Cosmochim Acta 1992;56(1):5−20.

[8] Roedder E. Studies of fluid inclusions; Part 1, Low temperature application of a dual-purpose freezing and heating stage. Econ Geol 1962;57(7):1045−61.

[9] Roedder E. Studies of fluid inclusions; [Part] 2, Freezing data and their interpretation. Econ Geol 1963;58(2):167−211.

[10] Su J. Study on geochemical characteristics of fluids of Badi copper deposits in Shangri-La. Kunming University of Science and Technology; 2014.

[11] Pan Z. Crystallography and mineralogy, Vol. 1. Beijing: Geological Publishing House; 1985.
[12] Wilkins RWT, Barkas JP. Fluid inclusions, deformation and recrystallization in granite tectonites. Contrib Mineral Petrol 1978;vol. 65(3):293–9.
[13] Roedder E, Belkin HE. Application of studies of fluid inclusions in Permian Salado salt, New Mexico, to problems of siting the Waste Isolation Pilot Plant [M]//Scientific basis for nuclear waste management. Springer US; 1979. p. 313–21.
[14] Swanenberg HEC. Fluid inclusions in high-grade metamorphic rocks from SW Norway. 1980.

Chapter 2

Classification of fluid inclusions

Chapter outline

2.1 Genetic classification of inclusions 19
2.1.1 Primary inclusions 19
2.1.2 Pseudo-secondary inclusions 20
2.1.3 Secondary inclusions 20
2.1.4 Metamorphosed inclusions 20
2.2 Classification of physical phase states of inclusions 21
2.2.1 Fluid inclusions 21
2.2.1.1 Pure liquid inclusions 21
2.2.1.2 Pure gas inclusions 21
2.2.1.3 Liquid inclusions 21
2.2.1.4 Gas inclusions 21
2.2.1.5 Daughter mineral inclusions 22
2.2.1.6 Carbon dioxide inclusions including liquid 22
2.2.1.7 Oil–gas inclusions 22
2.2.2 Magmatic inclusions 23
2.2.2.1 Crystalline melt inclusions 23
2.2.2.2 Vitreous melt inclusions 23
2.2.2.3 Fluid melt inclusions 23
References 24

Classification of inclusions is an important research topic in geology. In mineral processing studies, the chemical composition of the inclusions is the focus of research. A brief explanation of the classification of inclusions is necessary as the basis of inclusion study. Inclusions can be classified by different methods. According to the cause of formation, they are divided into primary, secondary, pseudo-secondary, and metamorphosed inclusions. Based on the physical phase state, inclusions are also classified as hydrothermal solution inclusions and molten silicate inclusions at room temperature.

According to the properties of the captured fluids during the formation of inclusions, there are inclusions captured from uniform (homogeneous) and nonuniform (heterogeneous) fluids. As mentioned before, most are normal inclusions of uniform fluids. A few abnormal inclusions are from nonuniform fluids. Table 2.1 shows the genetic classification and state classification of fluid inclusions at room temperature.

Fluid Inclusion Effect in Flotation of Sulfide Minerals. https://doi.org/10.1016/B978-0-12-819845-2.00002-8

TABLE 2.1 Classification of inclusions.

		No. of phases	Inclusion types from uniform and nonuniform fluids	Brief description
Physical state classification	Hydrothermal solution inclusions	1	Pure liquid inclusions	All liquid phase at room temperature
		1	Pure gas inclusions	All gaseous phase at room temperature
		2	Liquid-rich inclusions	Liquid phase accounting for more than 50% of total volume
		2	Gas-rich inclusions	Gas accounting for more than 50% of total volume
		≥3	Inclusions containing minerals	Including primary, gaseous phase, and daughter minerals
		≥3	Liquid CO_2 inclusions	Liquid, gaseous phase CO_2, and aqueous solution below critical temperature of CO_2
		≥3	Organic inclusions	Including organic liquid, gas, or solid other than liquid and gaseous phases
	Molten silicate inclusions	1 or ≥2	Amorphous melt inclusions (or glass inclusions)	It can be single-phase glass or made of glassiness and bubble
		≥2	Crystalline melt inclusions	Crystalline and bubble, sometimes with daughter minerals
		≥3	Molten solution inclusions	Crystalline, bubble, and solution, sometimes with daughter minerals
Genetic classification			Primary inclusions	Formed simultaneous with host crystals
			Secondary inclusions	Formed after host crystal and distributed in the fissures of the host crystal; cutting through the grain of the host crystal
			Pseudo-secondary inclusions	Formed after host crystal and distributed in the fissures of the host crystal
			Metamorphosed inclusions	Formed by metamorphism

According to Zhang W, Chen Z. Fluid inclusions geology. Wuhan: China University of Geosciences Press; 1993.

2.1 Genetic classification of inclusions

In geological studies, the genetic type of the fluid in inclusions is critical for interpreting the pressure, temperature, volume, and component data of the inclusions. By analyzing the differences between the inclusions and the host crystals, the inclusions can be divided into secondary, primary, pseudo-secondary, and metamorphosed inclusions. Their details are described next.

2.1.1 Primary inclusions

"Primary inclusions" refers to the inclusions formed by crystallization or recrystallization during the formation of the host crystal in naturally occurring minerals. Uniform or nonuniform fluids are captured in the formation process (see Fig. 2.1). Primary inclusions and the host crystal are formed in the same period. In other words, the composition of the primary inclusions is the same as the composition of the host crystal; primary inclusions are randomly generated and distributed in the host crystal, representing the mineral composition of the host crystal before mineralization. Figs. 2.1 and 2.2 show the primary inclusions along the crystal growth plane (see location *a* in Fig. 2.1) and zones, respectively. From the inside to the outside of the crystal,

FIGURE 2.1 Inclusions in quartz crystals. *a*, Primary inclusions; *b*, pseudo-secondary inclusions; *c*, secondary inclusions. *According to Lu H, Fan H, Ni P. Fluid inclusions. Beijing: Science Press; 2004.*

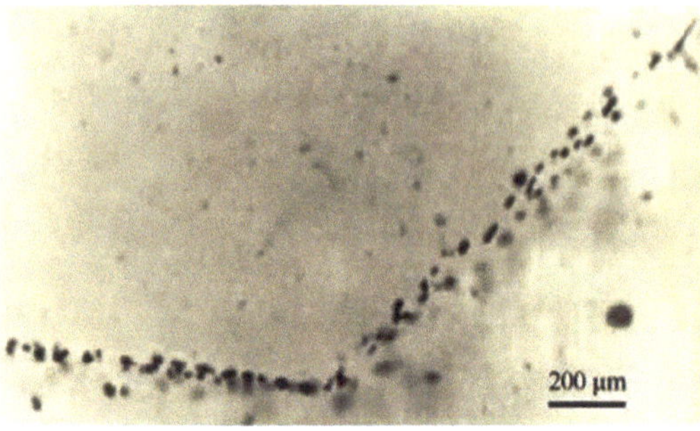

FIGURE 2.2 Primary inclusions distributed along the crystal growth zone. *According to Huanzhang Lu, Problems of Fluid Inclusions Petrography [J], Journal of Geological Geology, 2014, 20(2): 177–184.*

the primary inclusions at different locations may have different homogenization and formation temperatures.

2.1.2 Pseudo-secondary inclusions

Pseudo-secondary inclusions are formed following this process. The host crystals develop fissures in the process of crystallization, which become filled with the ore-forming fluids. After the fissure heals, the ore-forming fluid is stored in the host crystal to form inclusions (Fig. 2.1B). Distributed along the fissures, pseudo-secondary inclusions share similar characteristics with secondary inclusions. Analysis has shown that pseudo-secondary inclusions and host crystals are formed at the same stage. Therefore, pseudo-secondary inclusions represent the composition of the host crystal.

2.1.3 Secondary inclusions

Secondary inclusions are formed following this process. After the mineral formation, the hydrothermal solution enters the host crystal through the cracked surface or fissures. The minerals are dissolved and recrystallized to form secondary inclusions (see Fig. 2.1C). Distributed along the fissures of minerals, secondary inclusions represent the fluids entering the host crystal at a specific time, rather than the ore-forming fluid of the host crystal.

2.1.4 Metamorphosed inclusions

Metamorphosed inclusions are formed during the metamorphism of host crystals. The metamorphosed inclusions require a long period of time and variable conditions and environment for their formation. Therefore, in-depth research contributes to the understanding of the main process of metamorphism.

2.2 Classification of physical phase states of inclusions

Inclusions can be divided into molten and fluid inclusions according to different phases of matter. "Melt inclusions" refers to molten silicate inclusions such as amorphous glass or crystalline melt inclusions. The main component is the molten silicate mass and the chemical composition has little effect on mineral flotation. Here, we focus on fluid inclusions.

2.2.1 Fluid inclusions

2.2.1.1 *Pure liquid inclusions*

Pure liquid inclusions are the inclusions containing only liquid content. Due to the low temperature during their formation process, it is necessary to cool the fluid inclusions for observation under low-temperature conditions.

2.2.1.2 *Pure gas inclusions*

Pure gas inclusions are the inclusions with a single-phase gas inside. When the inner condensed gas is cooled, there will be a liquid phase at the edge of the inclusion. For noncondensed gas, the small liquid-phase circle cannot be seen. These inclusions are generally formed under high-temperature conditions, such as volcanic eruption and boiling.

2.2.1.3 *Liquid inclusions*

The liquid inclusion is a two-phase inclusion consisting of a liquid phase and a small bubble. Liquid inclusions are the most common and most widely distributed inclusions. Generally, the filling degree is used to characterize the volume percentage of the liquid phase:

$$\text{Filling Degree} = (V_L/V_{total}) \times 100\%,$$

where $V_{total} = V_L + V_G$; V_L is the volume of the liquid phase in the inclusions, V_G the volume of the gaseous phase, and V_{total} the total volume of the inclusions. In geology, "liquid inclusions" refers to the inclusions with a filling degree of more than 60%, while "gas inclusions" refers to the inclusions with a filling degree of less than 50%. However, this is only a statistical concept, without strict meaning. Liquid inclusions are the inclusions with a large filling degree (high fluid density) at room temperature and which are homogenized to liquid phase at high temperature. For example, in pure water inclusions, the density (or filling degree) should be greater than the critical density of H_2O (0.319 g/cm^3); similarly, the density of pure CO_2 inclusions should be greater than the critical density of CO_2 (0.46 g/cm^3).

2.2.1.4 *Gas inclusions*

Gas inclusions are inclusions filled with gases and with little liquid inside, which when heated is reversed to the uniform gaseous phase. Gas inclusions are common in melt hydrothermal and porphyry deposits.

2.2.1.5 Daughter mineral inclusions

Daughter mineral inclusions are the inclusions consisting of gases, liquid phase, and daughter minerals. The common states of daughter minerals are potassium salts, fluorite, hematite, and calcite.

2.2.1.6 Carbon dioxide inclusions including liquid

These inclusions are mainly composed of saline solution and carbon dioxide. During the heating process, the gas and liquid phases of carbon dioxide occasionally have a critical uniformity at 31.1°C, thus forming a single carbon dioxide phase.

2.2.1.7 Oil–gas inclusions

Oil–gas inclusions contain a gas–liquid multiphase and hydrocarbons. Fig. 2.3 shows four common oil–gas inclusions in minerals, namely gas–liquid aqueous, H_2O–CO_2, daughter mineral, and oil–gas inclusions.

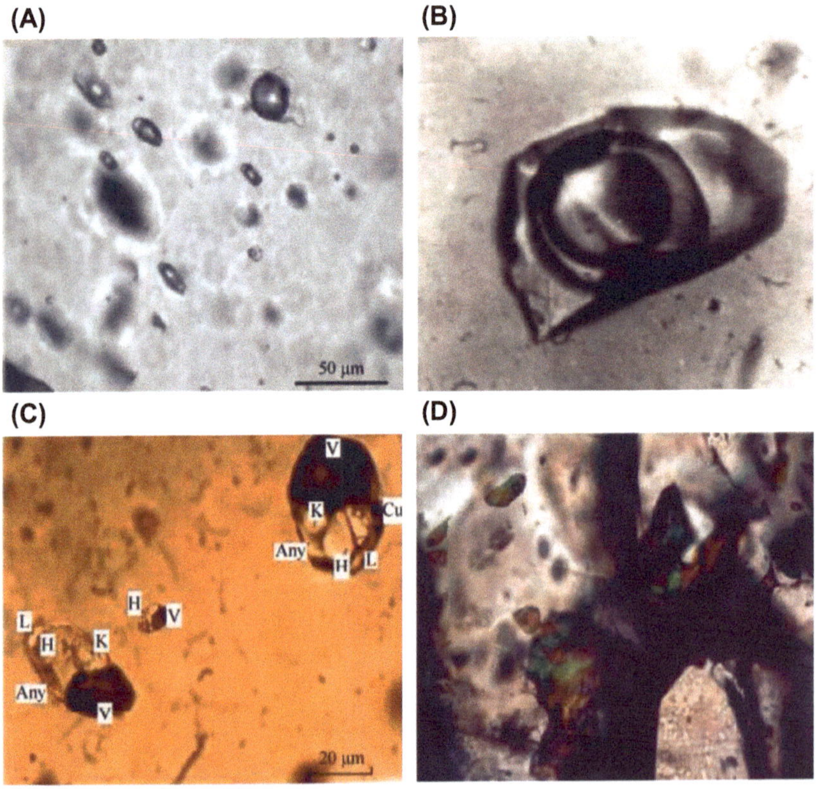

V-vapor phase; L-liquid phase; H-halite; K-sylvite; Any- anhydrite; Cu- Cu sulfides

FIGURE 2.3 Common inclusions in minerals. (A) Gas–liquid aqueous inclusions. (B) H_2O–CO_2 inclusions. (C) Daughter mineral inclusions. (D) Oil–gas inclusions. *According to Lu H, Fan H, Ni P. Fluid inclusions. Beijing: Science Press; 2004.*

2.2.2 Magmatic inclusions

"Melt inclusions" refers to the inclusions generated from captured molten materials during the growth of molten crystalline minerals. Compared with fluid inclusions, melt inclusions are yet to be fully explored in detail. Melt inclusions are divided into three subgroups.

2.2.2.1 Crystalline melt inclusions

Crystalline melt inclusions, common in intrusive rocks, are formed by the crystallization of captured molten silicate mass during a slow cooling process. This kind of inclusion consists of a bubble and crystallized minerals from the mass of molten silicate in the mineral.

2.2.2.2 Vitreous melt inclusions

Vitreous melt inclusions are composed of uncrystallized glass and a bubble. They are formed by the rapid cooling of magma at high temperatures. Vitreous inclusions are common among such inclusions, where the total content is not larger than half the total volume.

2.2.2.3 Fluid melt inclusions

The fluid melt inclusions are the fluid phase between the melted crystalline mineral of silicate and the gaseous phase, which are mainly separated from magma. For example, inclusions with concomitant molten silicate and sulfide beads are found in basic and ultrabasic olivine crystals.

Some examples of melt inclusions discovered are outlined as follows. Li Su et al. found some captured primary melt inclusions in the rock-forming mineral olivine orthopyroxene spinel of pine groves in the Qinling orogenic belt (see Fig. 2.4A–C). These inclusions are mostly isolated. The small inclusions produced by crystallization shrinkage are distributed around large inclusions (see Fig. 2.4A). Most of the individual melt inclusions are irregular in shape, with diameters mainly less than 10 mm (occasionally up to 25 mm); they have complex internal phase compositions, usually containing multiple phases of daughter minerals, such as the phases of silicate and opaque metal mineral (see Fig. 2.4A–C). In the individual inclusions, H_2O and CO_2 are found by laser Raman probes.

Bin Zhao et al. also found massive melt and fluid–melt inclusions in addition to gas–liquid inclusions in skarn minerals of Fe, Cu (Au), and Au (Cu) deposits along the Yangtze River of Daye-Jiujiang. These inclusions are found with various shapes, and are composed of a crystalline silicate (CSi), iron (Fe), amorphous silicate (ASi), and gas (V) multiphase of different phase ratios. Some of the inclusions contain several phases of crystalline silicate. The sizes of melt inclusions are generally (10–46) × (6–15) μm. The difference between a fluid–melt and a melt inclusion is that the former has a liquid phase and low uniform temperature.

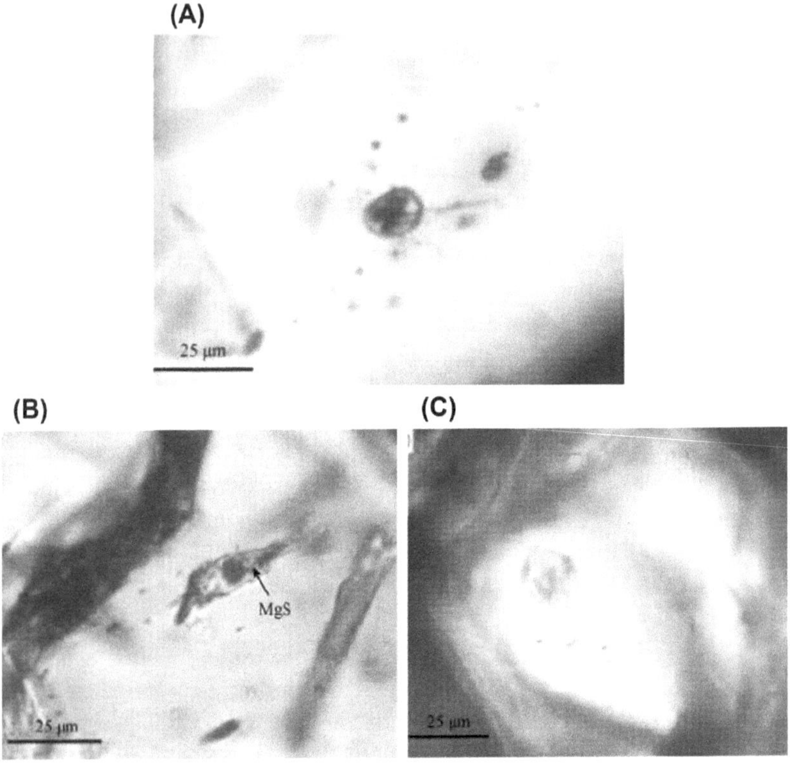

FIGURE 2.4 Micrographs of pure olivine rock mass in Qinling pine groves. (A) Oval-shaped melt inclusions in harzburgite (surrounded by small inclusions). (B) Multiphase melt inclusions in the coarse olivine residue of mylonite. (C) Negative crystalline melt inclusions in striplike ferrochrome chromite dunite. *According to Su L, Song S, Zhou D. Genesis of pine groove dunite in qinling orogenic belt: constraints of geochemistry and melt inclusions. Sci. China: D 2005;35(1): 38–47.*

References

[1] Zhang W, Chen Z. Fluid inclusions geology. Wuhan: China University of Geosciences Press; 1993.

[2] Lu H, Fan H, Ni P. Fluid inclusions. Beijing: Science Press; 2004.

[3] E R. Fluid inclusions evidence for immiscibility in magmatic differentiation. Geochem Cosmochim Acta 1992;56(1):5–20.

[4] Lu H. Problems of fluid inclusions petrography. J Geol Geol 2014;20(2):177–84.

[5] A V S. Melt inclusions in minerals as a source of principle petrological information. Petrology 1996;4(3):209–20.

[6] Su L, Song S, Zhou D. Genesis of pine groove dunite in qinling orogenic belt: constraints of geochemistry and melt inclusions. Sci. China 2005;35(1):38–47.

[7] Yang X, Peng L, Dong X. Melt inclusions characteristics and geological significance of Jinchuan Cu-Ni superbasic rock mass. Geol Rev 1991;37(1):70–9.

[8] Zhao B, Li Z. Characteristics of melt inclusions in skarn minerals of Fe, Cu (Au) and Au (Cu) deposits in Daye-Jiujiang area. Sci China 2002;32(7):550–61.
[9] Bodnar RJ. Introduction to fluid inclusions. Fluid inclusions: analysis and interpretation, vol. 32; 2003. p. 1–8.
[10] Austein RH. Petrographic analysis of fluid inclusions. Fluid inclusions: analysis and interpretation, vol. 32; 2003. p. 9–53.
[11] Shepherd TJ, Rankin AH, Alderton DHM. A practical guide to fluid inclusions studies. Blackie; 1985.

Chapter 3

Methods for the detection and composition study of fluid inclusions

Chapter outline

3.1 General optical microscopy research on fluid inclusions 29
3.1.1 Preparation of inclusion sheets 29
3.1.1.1 General requirements for the production of double-sided polished sheets 29
3.1.1.2 Grinding process of double-sided polished sheets 31
3.1.2 Identification of fluid inclusions 31
3.1.3 Morphology of fluid inclusions under the microscope 33
3.1.3.1 Shape 33
3.1.3.2 Size 34
3.1.3.3 Distribution 35
3.1.3.4 Abundance 36
3.1.3.5 Color 37
3.1.4 Phase state of matter in fluid inclusions 38
3.1.4.1 Gaseous phase 38
3.1.4.2 Liquid phase 38
3.1.4.3 Solid phase 38
3.1.4.4 Hydrocarbon 38
3.2 Modern research techniques for identifying fluid inclusions 39
3.2.1 Ultraviolet light microscopy 39
3.2.2 Infrared microscopy 39
3.2.3 Scanning electron microscopy 41
3.2.4 High-resolution X-ray computed tomography 44
3.3 Determination of inclusions' salinity 44
3.3.1 Basic principle of determining salinity of inclusions by the freezing method 45
3.3.2 Notes on the determination of salinity in inclusions by the freezing method 46
3.3.3 Instruments for measuring the freezing point 47
3.4 Extraction and analysis of fluid inclusion components 48
3.4.1 Issues to be aware of in composition analysis of inclusions 49
3.4.2 Analysis of chemical composition of group inclusions 50
3.4.2.1 Extraction of inclusion component fluids 50
3.4.2.2 Component analysis methods for liquid and gas in inclusions 52

Fluid Inclusion Effect in Flotation of Sulfide Minerals. https://doi.org/10.1016/B978-0-12-819845-2.00003-X

3.4.3 Analysis of the chemical composition of individual inclusions 57
3.4.3.1 Destructive analysis of individual fluid inclusions 57
3.4.3.2 Nondestructive analysis of individual fluid inclusions 61
References **66**
Further reading **67**

The detection and chemical composition study of fluid inclusions in minerals are two important components of inclusion research. In this chapter, we explain the existing methods of inclusion detection and chemical composition analysis. The procedure involves the detection and imaging of the fluid inclusions inside the mineral. After that, the morphology, which includes the size, shape, phase, distribution, and abundance, of the fluid inclusions is analyzed. Second, the internal material composition of the inclusions is discussed. Inclusion morphology detection methods include ordinary optical microscopy, infrared–ultraviolet optical imaging, scanning electron microscopy (SEM) analysis, and high-resolution X-ray tomography. The ordinary optical microscope is used for detection and morphology research of transparent and translucent minerals, excluding the opaque ores such as sulfides, under visible light. In 1984, Campbell et al. [16] used a near-infrared light as a light source to achieve the infrared optical imaging of the internal structures and fluid inclusions of opaque minerals. Later, a series of opaque–translucent mineral fluid inclusions was analyzed, including pyrite, wolframite, sulfur arsenic, stibnite, hematite, and hausmannite. This provided a reference for the research on opaque mineral inclusions.

The research on the composition of inclusions involves the mass and individual inclusions in minerals. The research on the composition of group inclusions in minerals generally includes single mineral selection, cleaning, opening of the inclusions and extraction, and analysis of the liquid components of the inclusions. The opening of the inclusions can be done by mechanical crushing, grinding, and thermal explosion methods. The thermal explosion method is often applied in geology, whereas the mechanical crushing and grinding methods are used in the mineral processing discipline. These two methods are closer to the grinding process, with more representative and statistical significance. The constants and trace elements present in fluid inclusions can be determined by ion chromatography (IC), atomic absorption spectroscopy, ICP-AES (inductively coupled plasma light), and ICP-MS (inductively coupled plasma mass spectrometry). After a vacuum thermal explosion of the mineral fluid inclusions, the extracted gaseous-phase components are determined by QMS (quadrupole mass spectrometry), GC (gas chromatography), or GC-MS (gas chromatography–mass spectrometry). The composition research on individual inclusions can be divided into destructive and nondestructive analysis. LA-ICP-MS (laser ablation inductively coupled plasma mass spectrometry) is the most widely used method for analysis of the liquid components of individual fluid inclusions. Other methods are SEM/EDS (energy dispersive X-ray spectroscopy) and SIMS (secondary ion mass spectrometry). The nondestructive methods for analysis of the internal components of inclusions include LRM (microscopic laser Raman spectroscopy), FTIR (Fourier transform infrared spectroscopy), and SXRF (synchrotron X-ray fluorescence), among others.

3.1 General optical microscopy research on fluid inclusions

In the research on fluid inclusions, the first step is to grind the mineral into slices or polished sections for preliminary morphological analysis under a transmission light electron microscope. Transmission light electron microscopy is still one of the most effective methods, as of this writing, for research on fluid inclusions. Optical microscopy, such as polarized light microscopy, can be used to analyze the morphology of fluid inclusions in transparent and translucent minerals, such as quartz and sphalerite. Polarizing microscopy identifies the single refraction or birefringence of a substance by changing the polarization of ordinary light. Any material with birefringence can be clearly imaged under a polarizing microscope, since birefringence is a fundamental feature of all crystals. Therefore, polarizing microscopes are widely used in research fields such as minerals, polymers, fibers, glass, semiconductors, and chemistry.

3.1.1 Preparation of inclusion sheets

Inclusions can be seen under the microscope through one of these sample methods of preparation: particle slides, slices, and double-sided polished sheets. The polished sheets are the most popular. Table 3.1 compares the main advantages and disadvantages of the three types of sheets.

The selection and production of inclusion sheets directly affect the petrographic research factors, such as the shape, size, abundance, and distribution of the inclusions. The requirements and specific processes for making polished sheets are described as follows.

3.1.1.1 General requirements for the production of double-sided polished sheets

(1) A representative sample is selected for sheet production. To prevent the influence of nontarget minerals, the part to be studied must be separately isolated for grinding.

(2) The mineral is cut into slices along the crystal axis or the required specific direction.

TABLE 3.1 Advantages and disadvantages of thin, particle, and double-sided polished sheets in optical research.

Type	Advantages	Disadvantages
Thin sheets	Easy to obtain and prepare; easy to store; easy to conduct mineral identification of the main rock	Thin sheets cannot be used for heating table research. This is because the resin for bonding easily decomposes and blackens at high temperature and easily breaks even at low temperature. The large inclusions are also easily broken. Sometimes the abrasive or small mineral fragments are embedded in the same resin and inclusions. Low-temperature inclusions are prone to precipitation at high temperatures (>100°C).
Particle or cleavage sheets	No special equipment is required; field operation; fast sample observation; selected inclusions for crushing station research	Not for long-term storage; restrictions on cold and hot tables; special oil immersion is required. The relationship of the inclusions cannot be well distinguished in some individual stripped particles.
Polished sheets	Polished sheets are directly used for hot table research and maintained for a long time; larger inclusions are preserved	It is difficult to identify minerals by optical properties because they are thicker than thin sheets. Dark or opalescent samples are made into very thin sheets (<100 μm), which are very difficult to detect.

According to Zhang W, Chen Z. Fluid inclusion geology, Wuhan: China University of Geosciences Press; 1993.

(3) Double-sided polished sheets need to be polished to a high degree. The image definition of the inclusions is proportional to the polishing degree of that sheet.

(4) The uniformity and thickness must be in accordance with the requirements. The sheet has a uniform thickness (generally 0.05–0.2 mm). Poorly transparent minerals such as sphalerite, milky quartz, and garnet require a small thickness. Transparent minerals such as crystal, calcite, beryl, and quartz have a less stringent requirement of thickness.

(5) The size of the polished sheet is determined according to the research requirements (usually 44 × 20 mm). In minerals, the distribution of inclusions is not uniform. Large inclusions are contained only in specific minerals. Therefore, it is necessary to count the abundance of inclusions in large-area inclusion sheets.

(6) The inclusion sheet is pasted and loaded by increasing the temperature. The temperature is generally controlled below 80°C to prevent the inclusions in the mineral from bursting.

3.1.1.2 Grinding process of double-sided polished sheets

(1) Positioning: According to the requirements of the sheet, the appropriate position is selected in the mineral.
(2) Slicing: A blade is used to cut the mineral samples at a thickness of 3–4 mm in the proper direction and position.
(3) Coarse grinding: The sheets are initially ground with 180–220 emery paper.
(4) Fine grinding: After the initial grinding of the sheet, fine grinding is performed with M10 abrasive paper.
(5) Polishing: Chromium oxide or ammonium dichromate is used as the polishing medium on a polishing machine. Manual polishing should be done on a glass sheet. The final polishing procedure is performed on a piece of flannel or a cloth.
(6) Adhesive sheet: The polished surface of the sheet is adhered to the production glass with abienic balsam or rosin.
(7) Grinding: The exposed surface can be treated with the aforementioned grinding and polishing procedures. However, the sheet should have large thickness.
(8) Sheet unloading: First, the sheet is removed from the glass at a temperature of 80°C. After that, the binder is washed away with an organic reagent, and then the sheet is dried. The sheet will be heated later. At high temperatures, any reagents that are burned will damage the instrument, thus affecting observation. Therefore, the sheet surface must be cleaned.

3.1.2 Identification of fluid inclusions

It is easy to confuse the fluid inclusions in mineral slices with impurities such as spots, water droplets, and polishing pits on the sample. Problems often occur in microscopic research of single liquid or gaseous-phase inclusions. Bubbles float in inclusions containing gaseous and liquid phases, which can be used to distinguish impurities from inclusions. Table 3.2 shows the differences between inclusions and impurities.

The following problems should be considered when finding inclusions in minerals:

(1) Unweathered grains with good crystallinity and transparency are advantageous for the observation of inclusions. Particularly, light and colorless crystals are most suitable for observation.
(2) The host crystal is selected. The inclusions are contained in host crystals that are crystallized from fluid, according to the definition of inclusions. However, in opaque metal minerals it is difficult to observe the inclusions; therefore, in most research on inclusions transparent minerals, including quartz, topaz, fluorite, gypsum, calcite, and pyroxene, are mainly selected as host crystals. Inclusions in quartz are best developed. Inclusions can also be found in translucent metal minerals such as sphalerite and cinnabar.

TABLE 3.2 Differences between inclusions and impurities.

	Inclusion	Impurity
Shape	Regular inclusions have the same or similar shape as the mineral (or negative crystal)	The impurity is solid in most cases; its crystal morphology is quite different from that of the host crystal
Phase state	Mostly two phase	Mostly solid phase
Relationship with the host crystal wall	There is an obvious gaseous- or liquid-phase boundary between the host crystal and the solids (such as daughter minerals and hyaline) in the inclusions	Impurity is usually closely associated with the host crystal
Formation time	Most daughter minerals such as salt have longer crystallization time than their host mineral; some daughter minerals can also be the same as the host crystal	Impurity has the same crystal system as the host crystal, but its crystallization is earlier than that of the host crystal

According to Lu H, Fan H, Ni Pei fluid inclusion, Beijing: Science Press; 2004.

(3) For observation, the microscope's magnification is adjusted from low to high. It is generally observed that under low power some small black spots in the mineral have a regular arrangement or strip-like distribution. On this basis, a higher-power microscope is used to observe these small black spots in the field of view. Fig. 3.1 shows the shape and details of the inclusions observed.

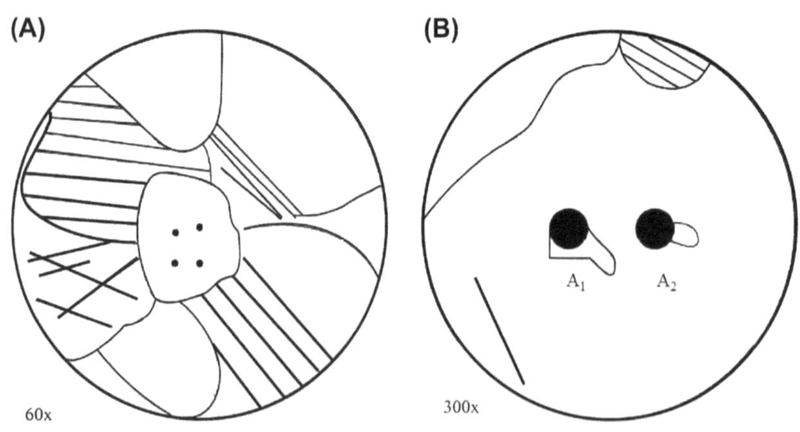

FIGURE 3.1 Positioning and identification of inclusions. (A) Positions of identified inclusions are marked with blue ink dots. (B) Details of inclusions are observed under a high-power microscope. A sketch of the inclusions is drawn and numbered. *According to Lu H, Li B, Shen K. Inclusion geochemistry, Beijing: Geological Publishing House; 1990.*

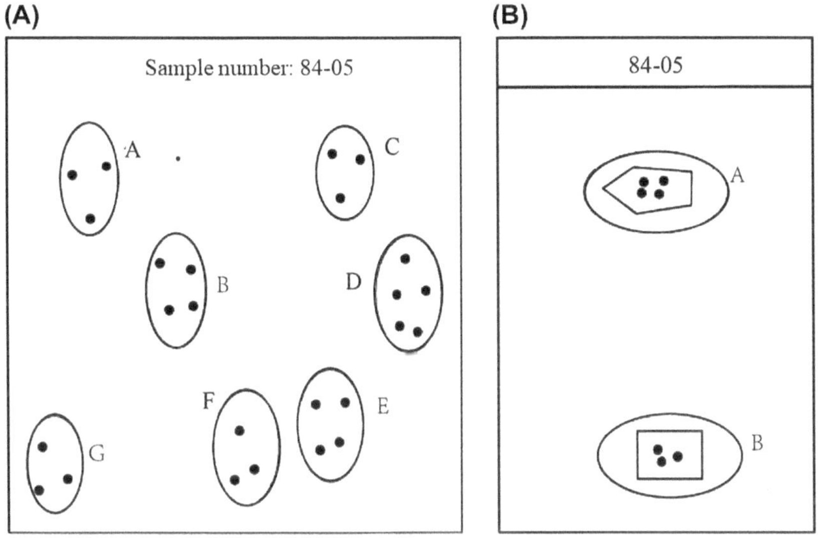

FIGURE 3.2 Positioning of inclusions. (A) Three to five fields of vision are selected for observation and numbering. (B) Identified and numbered sheets are cut into small pieces and placed in concave sheets. After numbering, the sheets are covered with flat glasses and fixed with horns. The sheets can be easily put into and taken from the transparent window area of the cooling and heating table. *According to Lu H, Li B, Shen K. Inclusion geochemistry, Beijing: Geological Publishing House; 1990.*

(4) The inclusions to be studied in depth are marked by positioning and numbering. To improve the test, features such as cleavage, cracks, and impurities around the inclusions are also marked (see Fig. 3.2).

3.1.3 Morphology of fluid inclusions under the microscope

Under a polarizing microscope, information on the fluid inclusions regarding the shape, size, color, quantity, occurrence, distribution characteristics, and filling degree can be obtained.

3.1.3.1 Shape

Common inclusions have regular or irregular shapes. Inclusions with regular shapes consist of negative crystalline inclusions (where the shape is similar to or the same as that of the host crystal), group inclusions with similar or the same shapes, and inclusions partially similar to the host crystal. The inclusions with irregular shapes consist of population inclusions (where the inclusions have different shapes) and inclusions with different shapes compared with the host crystal.

3.1.3.2 Size

The sizes of inclusions can be expressed as the long diameter measured by the micrometer under the microscope. Most inclusions under the microscope have long diameters of 2–20 μm. For a few inclusions, the long diameters reach hundreds of micrometers, with extremely rare diameters of more than 1 cm. The size limit of inclusions that can be observed by an electron microscope is 2×10^{-5} mm. The optimal observation range for inclusions is 10–100 μm. The research on inclusions focuses on their volume and size. For inclusions with regular shape, the length and width can also be measured under the microscope, while the depth of the body cavity is obtained by an estimation method. When the length, width, and height of the regular inclusions are known, the volume of inclusions can be determined by the volume equation. However, for inclusions with irregular shapes the dimensions are difficult to estimate by this method. In 1983, Bodnar described the volumetric estimation of irregular fluid inclusions. The principle is as follows:

It is assumed that the volume and mass of the inclusion are constant. Then, the total volume (V_1) is equal to the total mass (M_1) divided by the total density (D_1).

The equation is expressed as follows:

$$V_1 = \frac{M_1}{D_1} \tag{3.1}$$

At any temperature (T), the total mass (M_1) of an inclusion is equal to the sum of the masses of all the phases (i, ..., n) in the inclusion.

$$M_1 = M_i^T + M_j^T + \cdots + M_n^T \tag{3.2}$$

The mass of any phase in an inclusion is equal to the volume of the phase multiplied by the density at the same temperature:

$$M_i^T = \left(V_i^T\right)\left(d_i^T\right) \tag{3.3}$$

At the any given temperature, the volume of any phase (such as V_i^T) in the inclusion can be expressed by subtracting the sum of the volumes of the other phases from the total volume of the inclusion (V_1):

$$V_i^T = V_1 - \left(V_j^T + \cdots + V_n^T\right) \tag{3.4}$$

Combining Eq. (3.2) into Eq. (3.1):

$$V_1 = \frac{V_i^T d_i^T + V_j^T d_j^T + \cdots + V_n^T d_n^T}{D_1} \tag{3.5}$$

Combining Eq. (3.3) into Eq. (3.5):

$$V_1 = \frac{M_i^T + M_j^T + \cdots + M_n^T}{D_1} \tag{3.6}$$

Combining Eq. (3.4) into Eq. (3.6):

$$V_1 = \frac{d_i^T\left[V_1 - \left(V_j^T + \cdots + V_n^T\right)\right] + V_j^T d_j^T + \cdots + V_n^T d_n^T}{D_1}$$

$$= \frac{d_i^T V_1 - d_i^T\left(V_j^T + \cdots + V_n^T\right) + V_j^T d_j^T + \cdots + V_n^T d_n^T}{D_1}$$

$$V_1 D_1 = d_i^T V_1 - d_i^T\left(V_j^T + \cdots + V_n^T\right) + V_j^T d_j^T + \cdots + V_n^T d_n^T$$

$$V_1 D_1 - d_i^T V_1 = -d_i^T\left(V_j^T + \cdots + V_n^T\right) + V_j^T d_j^T + \cdots + V_n^T d_n^T$$

$$V_1\left(D_1 - d_i^T\right) = -d_i^T\left(V_j^T + \cdots + V_n^T\right) + V_j^T d_j^T + \cdots + V_n^T d_n^T$$

$$V_1 = \frac{-d_i^T\left(V_j^T + \cdots + V_n^T\right) + V_j^T d_j^T + \cdots + V_n^T d_n^T}{D_1 - d_i^T} \tag{3.7}$$

Eq. (3.7) is the basic equation for calculating the volume of fluid inclusions. If the solution in the inclusions is water, existing as a bubble, liquid, or vapor, then Eq. (3.7) is simplified into Eq. (3.8) for the two-phase pure water inclusions at 25°C:

$$V_1 = \frac{V_2 \quad d_1}{D_1 - d_1} \tag{3.8}$$

where the subscripts 1 and 2 are the liquid and gaseous phases (bubbles) corresponding to *i* and *j* in Eq. (3.7), respectively. T is 25°C in Eq. (3.7).

In Eq. (3.8), the gaseous-phase volume $V_2 = 1/6\pi D^3$ (where D is the diameter measured by ocular micrometer under a microscope). The densities of liquid water and water vapor at 25°C are 0.997 (d_1) and 0.000023 g/cm^3 (d_2), respectively. The gas–liquid inclusions are heated to a uniform temperature to find the isodensity line along the gas–liquid curve of the pressure–volume–temperature phase diagram of water.

3.1.3.3 Distribution

According to their aggregation state and arrangement, the inclusions in minerals are classified into regular and irregular distributions. Regular distribution refers to the phenomenon of regular arrangement of the group inclusions. It has been determined that the distribution of inclusions is restricted to the crystal growth mechanism of the host crystal and, in addition, in regular distribution, the arrangement has the same direction as the crystal plane, edge, or growth texture of the host crystal. There are several cases of regular distribution in inclusions, which are discussed next.

(1) The inclusions are distributed along the microcrystal plane of the host crystal in a ringlike pattern, which is similar to the ringlike structure of the mineral. (2) The inclusions are distributed in parallel strips. (3) The shapes of the individual inclusions are different. However, each inclusion has a regular geometric shape. In particular, the corresponding sides (certain crystal planes) of the inclusions are parallel to one another. (4) In general, group inclusions have individual inclusions with similar shapes and directional long axes. These inclusions are sketched to present a clear crystalline outline of the host crystal. (5) The individual inclusions have irregular shapes in group inclusions. However, the long-axis direction of each inclusion is the same as the crystal surface of the host crystal. (6) The individual inclusions have irregular shapes in the group inclusions. In addition, the long axes of the individual inclusions have an uncertain direction. Yet, the group inclusions stretch in a strip shape. In some particular cases, the strip extends in equal width, indicating that the group inclusions are evenly distributed over a set of parallel microcrystalline surfaces. Sometimes, the inclusions are distributed in lines; these inclusions may be secondary inclusions (or pseudo-secondary inclusions) with uneven distribution along the fissure. Irregular distribution of inclusions refers to the irregular distribution of group inclusions in the host crystal. Nevertheless, the shapes of the individual inclusions can be either regular or irregular.

3.1.3.4 Abundance

Generally, abundance is used to characterize the amount of fluid inclusions in the mineral. In a single crystal, the total volume of inclusion is typically less than 0.1% of the known crystal volume (see Fig. 3.3). Their abundance and distribution depend on the original growth conditions of the crystal and the history of its crystallization.

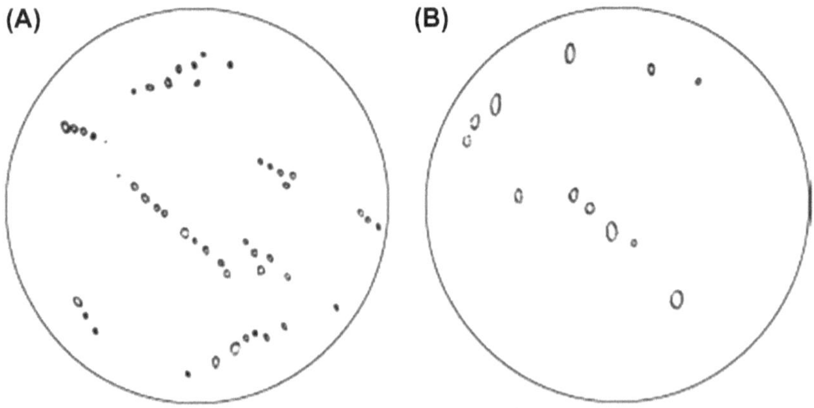

FIGURE 3.3 Typical fields of vision when the inclusions account for 1% of the sample (assumed volume percentage = area percentage). (A) vision field for small size inclusions (B) vision field for big size inclusions. *According to Zhang W, Chen Z. Fluid inclusion geology, Wuhan: China University of Geosciences Press; 1993.*

The numbers of inclusions in mineral crystals can be roughly estimated on the assumption that the distribution of inclusions is uniform, but in actual fact, the distribution of inclusions is mostly nonuniform. Many inclusions are formed in the early stage of mineral molding, and the number greatly diminishes at the late stage. For example, the near root of a quartz crystal is opaque due to the large number of inclusions, whereas the top is very transparent due to the small number of inclusions. For minerals with ribbons (such as fluorite and sphalerite), the internal ribbon has more inclusions than the external ribbon.

The number of fluid inclusions in the mineral can be counted under a microscope. First, the numbers in 1 cm^2 are counted to calculate the number in 1 cm^3. After determining the sizes of the inclusions and the number of inclusions per square centimeter, we calculate the total volume of the inclusions per cubic centimeter. Then, the total volume is compared with the volume of the host crystal to determine the ratio of the two. The number of inclusions can also be easily and roughly estimated. For example, when the average sizes (long diameter) of the inclusions are 1 mm, 100 μm, 10 μm, and 1 μm, and we assume the volume of the inclusions is 0.1% of the total volume of the sample, the average (estimated) numbers of inclusions present per cubic centimeter of the sample will be 1, 10^3, 10^6, and 10^9, respectively.

3.1.3.5 Color

The color of the inclusions is affected by the color of the internal substance and external factors such as the thickness of sheet and the refractive effect of light. The research observations of previous studies of fluid inclusions can be summarized as follows. The presence of a certain amount of hydrocarbons in the liquid phase of inclusions is generally characterized by yellow or brown appearance. In the presence of coloring ions in the liquid phase, the appearance of the inclusions is determined by the color of the ions. For example, the colors of inclusions are blue, light purple, light green, and light red when they contain Cu^{2+}, Fe^{2+}, Fe^{3+}, and Mn^{2+} ions, respectively. In the absence of colored ions, the brine solution and water vapor do not give any colored appearance in inclusions. The gas–liquid interface has a distinct annular black line and the center of the bubble is bright. However, the bubble may sometimes reflect a certain color due to light refraction. The bubbles are usually light brown and black when containing organic gas and carbon, respectively. False colors may be seen under a high-power microscope, and there may be no bright spots in the center of the bubble. Therefore, operators must be careful when observing the color of inclusions under a high-power microscope (600×–1000×). Unlike fluid inclusions, the color of melt inclusions is more complex. This is because the solid matter in melt inclusions has a specific color, which ultimately affects the color presented by the inclusions. When silicate glass is the main substance present in inclusions, the color of the inclusions is determined by the silicate glass. With the presence of a significant amount of Fe and Mg, the inclusions show dark green or green color, respectively.

3.1.4 Phase state of matter in fluid inclusions

The substances in mineral fluid inclusions can be divided into gaseous, liquid, and solid phases according to their form of existence. The most common form of inclusion is the gas–liquid type. There is also the single gas- or liquid-phase and multiphase inclusions. The phase states of the substances in the inclusions are described next.

3.1.4.1 Gaseous phase

When the gaseous and liquid phases coexist in inclusions, the gaseous phase generally appears as spherical bubbles. Bubbles continually colliding have been observed in some small inclusions, corroborating a gaseous phase in the inclusions. Due to thermal expansion and contraction, the bubbles are released within the system after heating, which changes the volume and increases the liquid phase. The bubbles in melt inclusions are usually brown or black, and generally, they do not appear as elliptical spheres. Sometimes, multiple bubbles generally appear as ellipsoid with high-viscosity gas. In single-gaseous-phase inclusions, the color is dark. There is a low light in the center of the gaseous phase, and a thick boundary with the host crystal. The microscope polarizer is pushed to cause insignificant change in the optical properties. If there is a change, the transparent position has the same interference color as the host crystal.

3.1.4.2 Liquid phase

In addition to bubbles, the liquid phase is the filling component of the inclusion cavity. Most fluid inclusions contain liquid phases with rich chemical components, liquid CO_2 or organic liquids. When there is no chromogenic ion in the gas–liquid inclusions, the liquid salt solution is mostly colorless and transparent, but when the component contains chromogenic ions, the liquid salt solution is mostly light purple and light blue. The organic liquids and liquid-phase CO_2 form three-phase inclusions with the gas and aqueous solutions.

3.1.4.3 Solid phase

The solid components contain vitreous, daughter mineral, and nonuniformly captured crystals or debris trapped inside the inclusions.

3.1.4.4 Hydrocarbon

Hydrocarbon fluid inclusions or petroleum inclusions are the original samples enclosed in mineral lattice defects or fissures during hydrocarbon migration and accumulation. The fluid is the mineral to be sought. Petroleum inclusions have similar properties compared with petroleum, such as color, phase, and fluorescence, and in inclusions, their color is affected by the density and composition of the hydrocarbons. Petroleum inclusions are transparent, opaque, or dark in color, with high asphaltene content. If the asphaltene content of the inclusions is low the color/appearance is lighter and more transparent. Condensate oil–gas inclusions are generally colorless and transparent, with gaseous–liquid phases. Pure gaseous hydrocarbon inclusions are rare, opaque, and convex, presenting a black-brown or gray-black appearance under transmitted light. In general, most hydrocarbon fluid inclusions are primarily found in sedimentary and epimetamorphic rocks, and their composition varies from pure methane to solid black asphalt. They are also found in geological environments from hydrothermal deposits to alkaline magmatic rocks.

3.2 Modern research techniques for identifying fluid inclusions

3.2.1 Ultraviolet light microscopy

The main targets of UV microscopy in inclusion research are substances that fluoresce under UV light, such as liquid hydrocarbons in petroleum inclusions. Based on this feature, the UV microscope can distinguish hydrocarbon liquids from aqueous solutions if the host crystal does not fluoresce. Different hydrocarbon components have different wavelengths under UV light, so their appearance in inclusions can be white, light blue, light green, or orange-yellow. The color of fluorescence changes from yellow and green to blue and orange, indicating that the intensity is weakened, until it eventually disappears. UV microscopy is seldom used in the research on fluid inclusions; however, its recent application using a spectrophotometer has greatly improved the measurement of individual inclusions.

3.2.2 Infrared microscopy

Infrared microscopy imaging is of great significance in the research on inclusions, particularly in opaque minerals. In fact, most metal sulfides, especially sulfides of Cu, Pb, and Zn, are opaque or translucent under optical microscopy. Ordinary optical microscopy is used only to analyze transparent minerals symbiotic with metal sulfide mineral. Campbell et al. in 1984 first used an infrared optical microscope to perform infrared optical imaging of fluid inclusions in metal minerals. Three years later, fluid inclusions were studied in thermodynamic feature research combined with infrared optical microscope and microscopic temperature analysis. In subsequent research, many researchers have successfully discussed opaque minerals in metallic sulfide ore hydrothermal deposits. The research on fluid inclusions of opaque metallic sulfide minerals in hydrothermal deposits is conducive to the research on physicochemical conditions and mineralization during formation. Fig. 3.4 shows the microscopic image of fluid inclusions in pyrite under an infrared light microscope.

The working principle of an infrared microscope is described here as follows. Infrared light with specific ranges of wavelengths is used to illuminate opaque minerals under visible light, making them transparent or translucent.

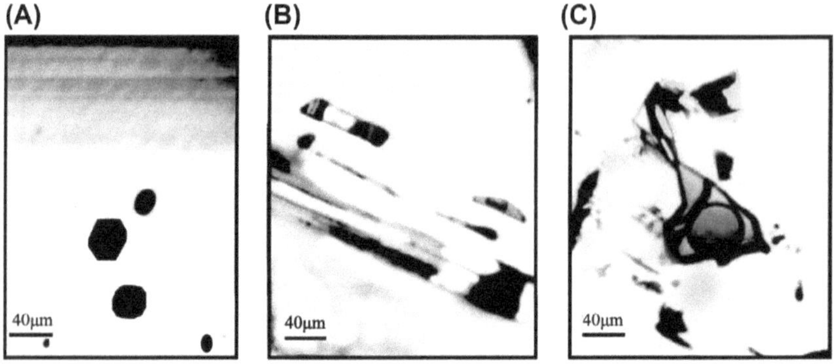

FIGURE 3.4 Images of fluid inclusions in an opaque mineral (pyrite) under infrared microscopy. (A) Equiaxed native inclusions parallel to the crystal plane (100) of pyrite. (B) Flat inclusions of a growth surface (100). (C) Large and flat inclusions in the center of a crystal. *According to Kouzmanov K, Bailly L, Ramboz C, et al. Morphology, origin and infrared microthermometry of fluid inclusions in pyrite from the Radka epithermal Cu deposit, Srednogorie zone. Bulgaria. Miner Depos 2002;37(6–7):599–613.*

The wavelength of the visible light is between 0.35 and 0.75 μm, and the quantum energy is between 5 and 1.65 eV. The boundary between the visible and the infrared light has a wavelength of 0.75 μm and energy of 1.65 eV. The boundary between the UV and the visible light has a wavelength of 0.35 μm and energy of 3.5 eV. In spectroscopy research on opaque minerals, the forbidden band determines the transparency degree of the mineral. The incident photon energy determines whether the photon is absorbed and whether the electron at the top of the valence band jumps to the conduction band. When the incident photon energy is larger than the absorption limit of the mineral, the mineral completely absorbs all visible light, which is why it appears as an opaque mineral. When the incident photon energy is less than the absorption limit of the mineral, visible light is transmitted through the mineral, and therefore, the mineral appears transparent. Fig. 3.5 shows the forbidden band levels of various minerals.

The infrared microscope system mainly consists of an object stage, infrared light source, collecting mirror, objective lens, focusing mechanism, image conversion tube, eyepiece, camera, and computer. The infrared microscope can provide a long wavelength of 2200 nm or less and a short wavelength of 1100 nm or less. A micro thermometer cold and hot table is widely used for fluid inclusions, with a temperature range of −195 to 600°C. The fluid inclusions observed under an infrared microscope are substantially identical to those observed under a polarizing microscope.

Infrared microscopy provides a new direction for fluid inclusion research. It is suitable for research on metal mineral fluid inclusions. In China, the research on mineral fluid inclusions under infrared microscopy is still an emerging

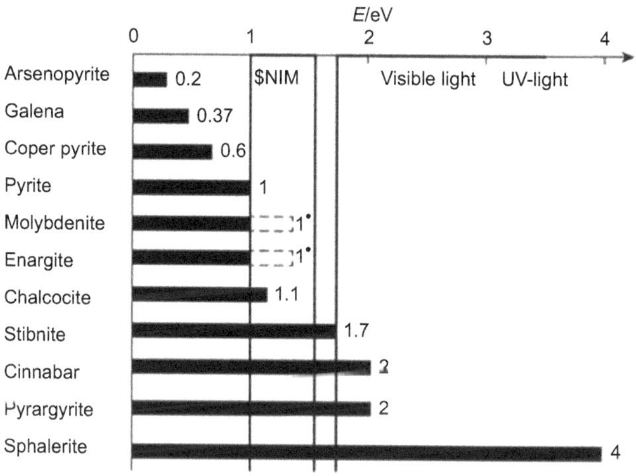

FIGURE 3.5 Forbidden band energy levels of common sulfides in hydrothermal deposits. *SNIM*, Scanning near infrared microscopy. *According to Shuey RT. Semiconducting ore minerals. Elsevier; 2012.*

technique with broad development prospects. However, some limitations are encountered in the research on fluid inclusions, particularly in opaque minerals.

(1) There are large differences between infrared transparencies of the same mineral. The mineralization factor of the mineral has a great influence on its transparency, and trace elements in the mineral also affect its transparency. (2) Under the infrared microscope, the imaging of fluid inclusions is not clear enough, even with large changes in transparency. The fluid inclusions in many metal minerals are black or dark gray, including the inclusions in pyrite and sphalerite. The phase states are difficult to distinguish. (3) The intensity of the infrared light affects the determination of salinity and type of inclusion. Under infrared light, the infrared filter cannot be used for inclusions, thus causing high temperature and deviation. (4) There are difficulties in the microscopic temperature measurement process. Under the infrared microscope, the image measurement of the inclusion is obtained from a data signal by infrared electronic induction. The infrared light refracts strongly when irradiating the inclusion wall, thus affecting the judgment of the ice crystal formation. The freezing point can be determined only by the changes in the size, shape, and position of bubbles during freezing and warming.

3.2.3 Scanning electron microscopy

SEM is one of the main testing instruments for in situ analysis of microareas. It has the advantages of high resolution, large depth of field, high magnification, and strong stereoscopic vision. Photographs taken by SEM have higher magnifications and clearer definitions than those by optical microscopy. SEM and EDS have been used to describe the quartz fluid inclusions of a Cu–Mo

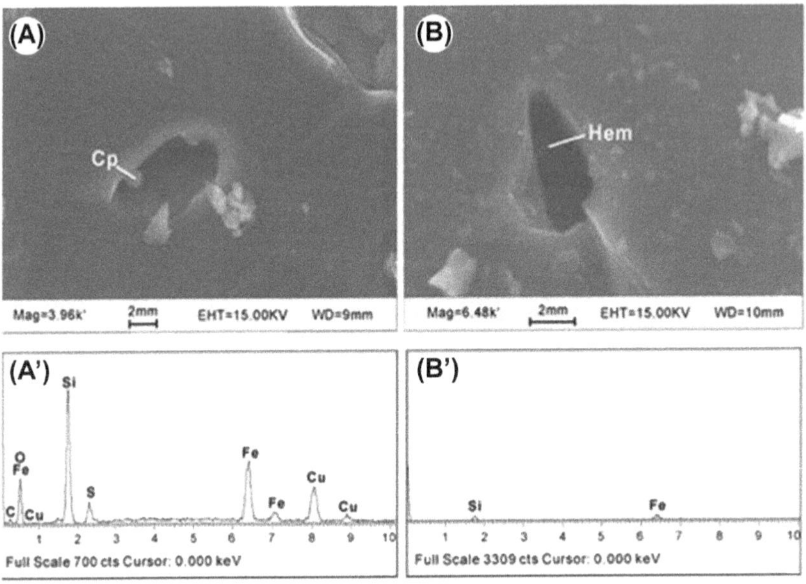

FIGURE 3.6 Scanning electron microscopy images and corresponding energy spectra of daughter minerals in quartz fluid inclusions of a Cu–Mo deposit in Inner Mongolia. (A and A′) Chalcopyrite (*Cp*) daughter minerals, (B and B′) hematite (*Hem*) daughter minerals. *According to Li N, Chen YJ, Ulrich T, et al. Fluid inclusions study of the Wunugetu Cu–Mo deposit. Inner Mongolia China. Miner Depos 2012;47(5):467–82.*

deposit in Inner Mongolia. Fig. 3.6 shows the SEM micrographs and energy spectra of quartz inclusions.

Early SEM was applied only to the morphology, location, and distribution of submicroscopic inclusions in the surface layer of fluid inclusions. However, SEM and EDS have great potential for identifying the "capture minerals" and daughter minerals in fluid inclusions. Even when the SEM is magnified 50×–20,000×, extremely clear photos or videos can still be obtained. Combined with EDS, SEM can be used to obtain a clear image that achieves a semiquantitative analysis of the daughter minerals in fluid inclusions. SEM is also suitable for analysis of both transparent and opaque daughter minerals.

Due to technical factors, SEM also has some defects in the research of fluid inclusions. For example, SEM can be used to measure only elements with atomic coefficients less than 5. Light sheets measured by a uniform method cannot be used for preselection under an optical microscope. It takes a long time to repeatedly search by X-ray the energy spectrum detection during the measurement process. In addition, SEM lacks the ability to distinguish between types of inclusions. It is difficult to determine whether a daughter mineral is a primary inclusion or secondary inclusion. Fluid inclusion samples are tested by SEM using the following procedure.

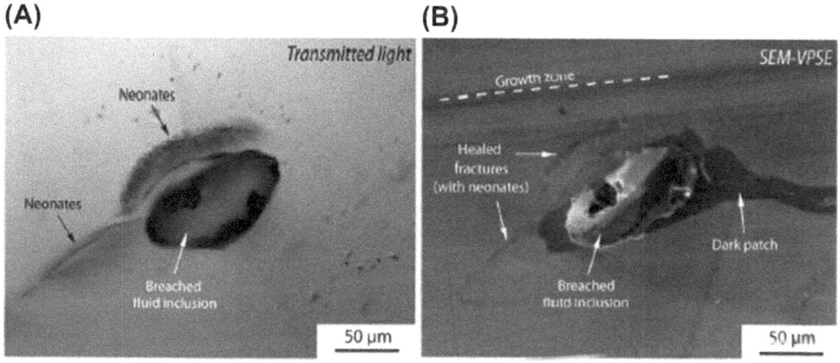

FIGURE 3.7 Micrographs of fluid inclusions in minerals. (A) Photomicrograph in transmitted light; (B) Scanning electron microscopy—variable pressure secondary electron (*SEM—VPSE*) imaging photograph at corresponding location. *According to Lambrecht G, Diamond LW. Morphological ripening of fluid inclusions and coupled zone-refining in quartz crystals revealed by cathodoluminescence imaging: implications for CL-petrography, fluid inclusions analysis and trace-element geothermometry. Geochem Cosmochim Acta 2014;141:381—406.*

First, the mineral inclusions in a light sheet are observed by an optical microscope to determine the quantity and characteristics of the inclusions and daughter minerals. The samples selected for SEM research (should have many daughter minerals and the complex types) are broken. Dozens of the fragments with diameters of about 1 mm are selected from the broken samples, and pasted on metal bases with diameters of 1 cm by a conductive adhesive. The fresh surface should be upward, that is, the surface of the fragment is parallel to the metal base to increase its surface area, thus improving the chances of finding substantial daughter minerals in the inclusion. The specimen is washed in an ultrasonic bath and then broken. The surface of the sample is coated with Au film using a sprayer. Finally, the sample coated with the Au film and the bases are used for comparative research by SEM. The samples are first scanned in proper order. Once a suspicious crystal is found in the cavity of the inclusions, it should be magnified and positioned (1000×—20,000×). The electron beam is excited by the X-ray detector to bombard the target daughter mineral, and the elements in the daughter mineral emit characteristic X-rays to obtain an intuitive spectrum in tens of seconds, which can be photographed. The supporting computer is used to obtain the relative data of each element.

Recently, foreign researchers have used SEM—VPSE (variable pressure secondary electron) technology to discuss mineral fluid inclusions. Fig. 3.7 shows the image of fluid inclusions by SEM—VPSE. This technique adopts the EVO 50 scanning electron microscope produced by Zeiss. Equipped with a VPSE detector, this microscope can be used to obtain better pictures of the morphology of fluid inclusions.

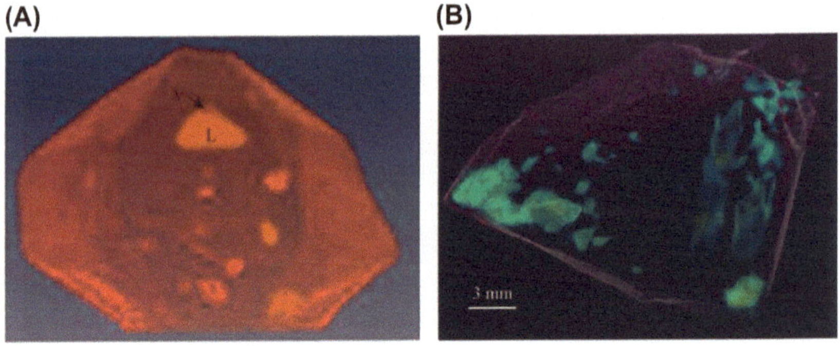

FIGURE 3.8 Three-dimensional images of fluid inclusions in (A) quartz and (B) sphalerite by high-resolution X-ray computed tomography. [8,9] *L*, liquid phase; *V*, vapor phase.

3.2.4 High-resolution X-ray computed tomography

High-resolution X-ray computed tomography (HRXCT) provides a multidimensional micrometer-resolution image of the internal structure of samples. HRXCT can scan thousands of particles at the same time. This unique technology brings great convenience to finding fluid inclusions with complete pores inside the mineral. Due to the large difference between the internal material compositions of the mineral inclusions and their host crystal, the X-ray signal reacts differently with the parts that contain fluid inclusions and the parts that contain no fluid inclusions in the mineral. Usually, after the HRXCT scanning, the shape and color of the spatial three-dimensional (3D) image region of the fluid inclusion change. Kyle et al. [8,9] used HRXCT to create a 3D diagram of the fluid inclusions in quartz crystal and sphalerite (see Fig. 3.8), showing that it is an effective method of distinguishing inclusions in opaque and translucent minerals. The advantage of X-ray tomography is that it does not rely on mineral transparency. The inclusions in the crystals are displayed in 3D form to find the distribution of both primary and secondary inclusions.

3.3 Determination of inclusions' salinity

In general, many fluid inclusions are rich in the chloride component. In geology, it is called salinity and is usually expressed as a percentage of NaCl. In pure mineral study, the presence of massive chloride ions in ore pulp solutions after grinding can be used as evidence for the release of fluid inclusion components. The salinity is indirectly determined with a microcalorimeter in individual fluid inclusions where the liquid component is mainly saline solution ($NaCl-H_2O$ or $NaCl-KCl-H_2O$ system). The salinity is determined by measuring the melting temperature of halite crystal in inclusions where the liquid component is mainly supersaturated brine. Salinity is also determined by measuring the freezing point on the hot and cold table by the freezing method in inclusions where the liquid component is mainly an unsaturated saline solution. The inclusions' salinity is determined based on the following theory.

There is a quantitative relationship between the eutectic point of the phase change of a brine solution and the electrolyte concentration of the solution at equilibrium in systems. Based on this quantitative relationship, the salinity of inclusions can be indirectly measured. Microcalorimeter results are often affected by optical properties in the measurement of water-rich inclusions; when the completely frozen inclusions are gradually heated to approach the melting point, the solid–liquid boundary between the inclusions and the host crystal will be blurred and difficult to distinguish; this is significant especially in small individual gas-rich inclusions. The microthermal method, on the other hand, exhibits large errors. In geology, the freezing method is generally used to determine the salinity of inclusions. First, the freezing point of the inclusions is obtained by microscopic temperature measurement of the individual fluid inclusions. After that, the freezing point is substituted into the empirical formula to calculate the salinity of the inclusions. The freezing point of the liquid phase in inclusions contributes to the acquisition of data related to mineralization information such as salinity, density, and ore-forming pressure. In addition, it is one of the important means to analyze the physical and chemical conditions of mineralization. For flotation in mineral processing, the chloride component in the inclusions has an important influence on the chemistry of the flotation solution and the adsorption of the double electrode layer on the mineral surface.

3.3.1 Basic principle of determining salinity of inclusions by the freezing method

The freezing method is an important method for fluid composition and salinity research on inclusions. The basic principle is described as follows. First, the temperature of the inclusions system is changed on the freezing table, causing a phase change in the inclusions with the temperature change. Compared with experimental phase diagrams of known fluid systems, we determine the system and composition of the fluid. For the $NaCl-H_2O$ inclusions with low salinity, the salinity of the fluid is determined according to Raoult's law, namely, the principle that the freezing point of a dilute solution decreases with the increase in solute molar concentration.

(1) For dilute solutions, the decrement of the freezing point of the solution is determined by the concentration of the aqueous solution rather than the type and nature of the solute. It can be expressed as:

$$\Delta t = k_f \times m \left(k_f = \frac{R \times (T_m)^2 \times M}{1000 \times \Delta H} \right) \tag{3.9}$$

where Δt is the decrement of the freezing point (°C), m the mass molar concentration (mol) of the solute, and k_f the constant for freezing point decrease. According to the k_f formula, k_f depends on the freezing point (T_m) of the pure solute, the molar mass (M) of the molecule, and the melting heat (ΔH). Therefore, it is relevant only to the nature of the solute. Various solutes of the same concentration have the same decrement of freezing point.

(2) The pressure has little effect on the decrement of the freezing point.
(3) Raoult's law is suitable for ideal solutions or nondielectric dilute solutions rather than strong dielectric solutions containing NaCl and small amounts of other salts. NaCl solutions with moderate concentration can be used under appropriate and modified conditions.

Hall et al. [17] derived a formula for calculating salinity using the freezing point decrement based on experimental data:

$$W = 0.00 + 1.78T_m - 0.442T_m^2 + 0.000557T_m^3 \tag{3.10}$$

where W is the weight percentage of NaCl and T_m the decrement of freezing point (°C).

3.3.2 Notes on the determination of salinity in inclusions by the freezing method

The salt concentration of inclusions can be calculated by measuring the freezing point of the solution according to Raoult's law, following the aforementioned freezing point drop formula or the actual brine system phase diagram. However, the following items should be noted when the concentration of the brine system in inclusions is measured by the freezing method:

(1) In general, the liquid-phase component of brine is a complex aqueous system in inclusions. Most components are dissolved in the NaCl solution, and it is impossible to determine the decrement of the freezing point caused by the solute components. The measured freezing point is caused by all the components in the inclusions, so the salinity obtained by the freezing method is actually a comprehensive result of multicomponent solutes, expressed as the concentration equivalent to NaCl. The existing experimental data show that the aqueous solutions of other major salts have similar freezing properties compared with the aqueous solution of NaCl. The freezing curves of KC1 and NaCl as expressed in moles are two nearly parallel lines, and NaCl is often the main component in solutions. Therefore, the freezing curve of an $NaCl-H_2O$ system can approximate the freezing curve of the multicomponent system.
(2) The freezing method is suitable only for the determination of $NaCl-H_2O$ systems with salinity of 0–23.3 wt%. If the salinity is out of this range, it cannot be measured by the freezing point. Therefore, the freezing point of an $NaCl-H_2O$ system cannot be lower than −21.2°C when the salinity is 23.3 wt%.

(3) In brine inclusions, the common systems include $KCl-H_2O$, $NaCl-KCl-H_2O$, $NaCl-CaCl_2-H_2O$, $NaCl-MgCl_2-H_2O$, $Na_2CO_3-H_2O$, $NaHCO_3-H_2O$, and $NaCl-H_2O$. If the fluid inclusions are composed mainly of components other than the $NaCl-H_2O$ system, then the salinity should be determined according to the corresponding experimental data of the system.

3.3.3 Instruments for measuring the freezing point

According to the principle of refrigeration, the freezing table can be divided into refrigerant and semiconductor cooling types. Actually, the two freezing tables have their applications. The refrigerant-cooling freezer is characterized by high freezing efficiency and low freezing temperature (as low as −190°C). This freezing mode also requires a special cooling system, special refrigerant, and complicated operation. The semiconductor freezing table can be directly cooled by electricity, without a special refrigerant-complicated operation. However, the freezing temperature can reach only −60°C, with low efficiency. With freezing and heating functions, the micro-heating table supplied on the market has a refrigerant and an electric furnace for low and medium–high temperature tests, respectively. Therefore, the same fluid inclusion can be tested at low and high temperatures to obtain salinity and uniform temperature. There are many manufacturers of cooling–heating tables in domestic and foreign markets. However, the common cooling–heating tables are mainly the THMSG600 produced by British Linkam Co. and USG by the US Geological Survey. Liquid nitrogen (boiling point −196°C) is the most common refrigerant in the laboratory; it is nontoxic, chemically inert, and nonflammable. In practical operation, the test should be completed in a ventilated room. When adding liquid nitrogen to the tank, the worker must be properly protected against liquid nitrogen splashing.

In 2001, X. Lu et al. applied Raman spectroscopy to determine the salinity of inclusions, and the results showed that the Raman spectrum is sensitive to changes in OH^- in aqueous solution in the range of 2800–3400 cm^{-1}. In addition, there are corresponding quantitative relationships between the concentration change in the electrolyte and the characteristic parameters of the Raman spectrum. The Raman spectral slope is used to reflect the solution concentration after analyzing the relationships between the Raman spectral characteristics and $NaCl-KCl$ aqueous solutions with different concentrations. The experimental analysis showed that there is a good linear relationship between the Raman spectral slope and the concentration of $NaCl-KCl$ in aqueous solution. A straight line was fitted to obtain the empirical formulas for calculating the salinity of $NaCl-KCl$ aqueous solutions by Raman parameters. Traditional microcalorimeters and laser Raman spectroscopy were used to measure the salinities of artificial and auriferous quartz vein inclusions, thus testing the validity of the empirical formulas. It demonstrated that laser Raman spectroscopy is an effective method to measure unsaturated fluid inclusions.

3.4 Extraction and analysis of fluid inclusion components

The origin of the analysis of and techniques for fluid inclusions in minerals began in the 1950s and became sophisticated in the 1970s. With the rapid development and integration of laser, optic, photoelectron, microelectronic, and computer technology, there have been many kinds of techniques developed for the determination of fluid inclusion composition and physicochemical properties since the 1980s. Meanwhile, modern testing techniques have been introduced into geological research, which make it possible to analyze high-precision components of inclusions. At this writing, the research on inclusions has entered a new phase. Inclusion composition analysis (including isotope analysis) has become an important part of inclusion research. It is one of the most important tasks in geology to establish the geochemical model of the process of ancient fluid action by providing the physicochemical information of ancient fluid components in fluid inclusions to the maximum extent. In mineral processing, the material composition and content of inclusions are equally important, involving a series of physicochemical effects of components on the mineral surface and the pulp system.

Up to now, the components of fluid inclusions have been analyzed by crush or burst extraction (also known as the colony method) or by direct methods of fluid extraction from individual fluid inclusions. Most fluid inclusions in minerals have long diameters of less than 50 μm and small volumes. Therefore, these two methods have their own advantages and disadvantages in practical applications.

The first method has the advantage of obtaining more fluid component samples. Multielement analysis can be performed in the same analytical procedure with massive fluid component samples. In addition, the concentrations of the detected elements in the fluids may be greater than the detection limit of most analytical instruments. However, this method is limited by many complex components and poor representativeness, leading to the low accuracy of the sample analysis, because massive fluid inclusions are composed of multiple generations in the same sample. Inclusions of different generations have different components. Multiple generations in the same sample lead to the complicated analysis of the inclusion colony components, thus increasing the final interpretation error in the results. This problem can be resolved by selecting representative samples of the same generation and a small volume.

For the second method, the biggest advantage is the control of the samples. The individual inclusions are accurately analyzed by microscopic observation to derive definite or unique geological information. The analysis of the individual inclusion results limits the number of elements that can be detected each time. However, this defect can be compensated for by rapid analysis and accurate sample control.

In mineral processing, researchers focus on the total amounts of components released by the inclusions rather than the generational problems of inclusions in the mineral. Thus the extraction of fluid inclusions by the group method is similar to the crushing and grinding of mineral processing engineering; therefore, the colony method is the focus of mineral processing.

3.4.1 Issues to be aware of in composition analysis of inclusions

High detection accuracy of inclusions can be obtained only by modern analytical technology. However, there are many problems in inclusion composition analysis due to the particularity of mineral inclusion samples.

(1) In the analysis of inclusion composition, especially when using the colony method, the primary issue is to eliminate the pollution and interference of the host crystal during the process of releasing the inclusion components. Mineral inclusions are complex systems that are trapped in the host crystal. In general, the volume is only a few thousandths of the host crystal or even smaller. Due to the small volumes and low internal component content, inclusions are easily interfered with by the host crystal component. Especially when the host crystal has a complex composition, it brings a nonnegligible influence to the analysis of the inclusion components. Therefore, it will be very effective to use individual minerals with high purity, to exclude the influence of other impurity.

(2) The volume of fluid in the inclusions is not more than dozens of micrometers, with few substances available for analysis. Such a small interval contains substances of solid, gaseous, and liquid phases. It is estimated that individual inclusions contain only about 10^{-9} mol of material, and the concentration of the important components (such as mineral-heavy metal elements) is lower (often less than 1 μg/g). Thus these factors increase the difficulty of the analysis and determine the diversity of the analytical methods.

(3) According to the statistics of existing research data, inclusions have extremely complicated components, which include a liquid phase that is a saline and hydrocarbon solution containing a large amount of solute. The concentration of the solute in the solution varies greatly (0%–60% NaCl or more). Present within are cations such as K^+, Na^+, Ca^+, Mg^{2+}, Mn^{2+}, Ba^{2+}, Al^{3+}, Si^{4+}, Mo^{2+}, Pb^{2+}, Zn^{2+}, Cu^{2+}, and Sr^+ and anions such as F^-, Cl^-, Br^-, HCO_3^-, SO_4^{2-}, HS^-, NH^-, and SO_2. Gaseous phases include H_2O, CO_2, CO, CH_4, C_2H_6, N_2, H_2, O_2, H_2S, and NH_2; solid phases include stone salt, potassium salt, calcium dichloride (Ca_4Cl_2), carbonate, feldspar, mica, pyrite, chalcopyrite, magnetite, ilmenite, rutile, pyroxene, and spinel.

(4) According to the different components, the appropriate analytical method is therefore selected to analyze the inclusion components. Many modern testing techniques can be used for inclusion analysis. However, some of these methods are suitable only for analyzing certain element or molecular

combinations. The high-precision test instruments have special effects on certain elements. For example, ICP is the best for measuring heavy metal ions, whereas the ion-selective electrodes are mainly used to analyze halogen elements.

(5) The component analysis must be closely integrated with the inclusion research, since different studies on inclusions require different sample test and preparation methods. For example, the gas components of primary inclusions are measured to identify the ore-forming fluid gas component in the host crystal formation period. The gas in secondary inclusions is excluded during the sample preparation. If the gas is extracted by vacuum gas explosion, then the gas in the secondary inclusions should be removed by explosion before the temperature rise in sections. Raising the temperature causes the primary inclusions to burst and release the gas for chromatographic analysis. Individual inclusion composition analysis can overcome the aforementioned shortcomings; however, the inclusions must be identified under the microscope in advance to select the required individual inclusions, thus performing the operation analysis.

3.4.2 Analysis of chemical composition of group inclusions

3.4.2.1 Extraction of inclusion component fluids

Sample preparation is the most critical preliminary task in inclusion analysis. Only by preparing representative samples and minimizing the contamination can better results be obtained from high-precision instrumental analysis. In the geological field, the process of preparing samples for analysis of group inclusions is integrated with the needs of inclusion research in mineral processing engineering disciplines. In composition research of group inclusions, the samples are generally prepared by individual mineral selection, cleaning, opening of the inclusions, and extraction of the component liquids.

3.4.2.1.1 Individual mineral selection

Minerals with high purity are used for composition analysis research of mineral group inclusions. When impurities are brought into the inclusion, they affect the quality of the analysis of the components, so the purity of the mineral intended for analysis should be as high as possible based on the existing technology. As of this writing, the common minerals used for inclusion analysis are quartz, fluorite, beryl, apatite, magnetite, garnet, galena, sphalerite, and barite; and quartz is the best option due to its wide distribution and lower pollution. Quartz has only a few impurities other than the well-known SiO_2. In addition, the inclusions are abundant and well preserved. In the selection of individual mineral, the mineral sample should be broken to the appropriate particle size. When a pure individual mineral is collected, then the final particle size is suitable for grinding or thermal explosion, but if fine mineral aggregate samples are collected, then the inclusions are broken for the purpose of monomer dissociation. The particle size should be large because a very small particle size will lead to the breakage and loss of inclusions during the crushing process. After breaking, the samples are sieved, washed, and dried (no more than 100°C) for purification.

3.4.2.1.2 Cleaning

The selected individual mineral sample is placed in a beaker and soaked in dilute hydrochloric acid for about 1 h (stabilized minerals such as quartz are recommended to be soaked in 1:1 dilute hydrochloric acid, while labile minerals such as sulfides and calcite are soaked in 1:4 (or less) dilute hydrochloric acid). During the soaking process, the solution is stirred to remove contaminants from the surface of the mineral particles or from the cracks. The sample is then poured into a glass sand funnel. After repeated washings with deionized water, the sample is filtered to achieve neutrality. Finally, an ultrasonic cleaner is used for vibration cleaning until the conductivity of the solution is close to that of the deionized water used for the sample cleaning. The clean sample is filtered again, and dried in an oven below 100°C. After that, the sample is stored and ready for use.

3.4.2.1.3 Opening of inclusions

At present, there are three methods for opening inclusions; they include mechanical crushing, grinding, and thermal explosion.

3.4.2.1.3.1 Mechanical crushing method The sample is placed in a stainless steel or copper tube closed at one end and with a vacuum valve at the other. The tube is then attached to a vacuum system and evacuated to 10^{-3} Pa. The valve is closed to remove the tube. After that, the sample tube is flattened with a press machine, breaking the mineral sample and opening the inclusions. With this method, the particle size of the broken sample cannot be smaller than 1 mm. Therefore, only the inclusions on fractured surfaces can be opened.

3.4.2.1.3.2 Grinding method (for mineral processing) Grinding is widely applied in mineral processing. There are two ways of grinding. (1) Agate mortar: the mineral sample and a little deionized water are placed in a pure agate mortar for artificial grinding. (2) Ceramic ball mill: the mineral sample and a little deionized water are put in the ceramic mill and ground with the ball. To open most inclusions, the ground particle size should be smaller than the diameter of the inclusions. The ground sample is thereafter centrifuged, and the clear liquid is transferred to a volumetric flask for component analysis.

3.4.2.1.3.3 Thermal explosion method The mineral sample is heated in a quartz tube until the sample bursts. In the thermal explosion method, the optimum sample particle size is 0.25–0.5 mm. The quantity of the inclusions can be determined from the temperature curve of the sample during the thermal explosion. The burst peak temperature of group inclusions is taken as the temperature at which the inclusions are released to determine the components. After the opening of the inclusions by thermal explosion, the volatiles are introduced into a quadrupole mass spectrometer or gas chromatograph for analysis. Soluble salts are also extracted from the inclusions to analyze the liquid-phase composition of the inclusions. The thermal explosion method has a larger amount of leaching than the grinding method. The mineral is ground into a fine powder with large surface area. The ions in the solution are inevitably adsorbed on the surface. If the ion concentration is constant in a generally neutral solution, then the smaller mineral particles will lead to adsorption of more ions. If the mineral particle size is constant, higher ion content in the solution will lead to less relative adsorption. The grinding method has the disadvantage of surface adsorption. When the mineral is finely ground, tiny mineral impurities and some alkalis entering the mineral lattice may be extracted together with the liquid-phase components of the inclusions. This results in errors in the analysis of the liquid phase. At present, the thermal explosion method is widely used to open inclusions for composition analysis of gas–liquid-phase domestic group inclusions.

3.4.2.1.4 Treatment before determination of liquid-phase component of inclusions

The ultrasonic water extraction centrifugation method is generally used in China. Experiments showed that this method can greatly reduce analytical error caused by ion adsorption due to grinding and the contamination of tiny mineral impurities. The salts in the opened inclusions can be completely extracted during this process. One to three grams of an inclusion sample opened by explosion is put into a quartz beaker. After that, deionized water (25 mL) is poured into the beaker for ultrasonic cleaning. The ultrasonic screen current is adjusted to 110 mA. Stable minerals such as quartz need ultrasonic cleaning for about 1 h, while labile samples such as sulfide require shorter time (about 40 min). Finally, the ultrasonic cleaning liquid is centrifuged. These steps are repeated four times. Each supernatant is removed into a 100-mL volumetric flask for ultrasonic vibration until the conductivity of the solution is close to that of the deionized water used for sample cleaning. The extracted supernatant is then concentrated for component analysis.

3.4.2.2 Component analysis methods for liquid and gas in inclusions

With the development of analytical detection technologies, more methods and instruments are used to analyze the trace components of fluid inclusions. In the extraction of inclusion components, the constants and trace elements can be

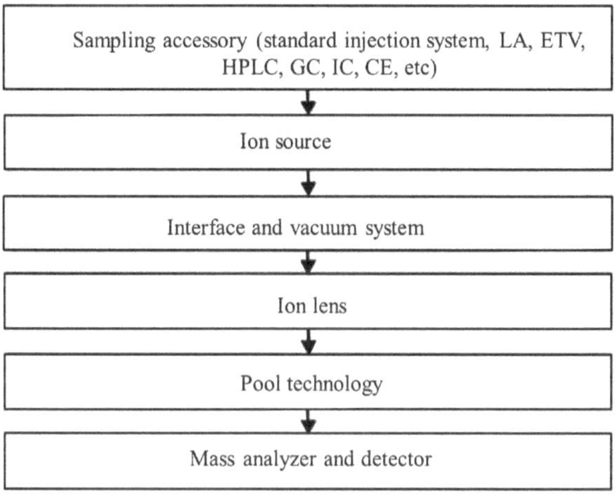

FIGURE 3.9 Inductively coupled plasma mass spectrometry hardware. *CE*, capillary electrophoreses; *ETV*, electro thermal vaporization; *GC*, gas chromatography; *HPLC*, high-performance liquid chromatography; *IC*, ion chromatography; *LA*, laser ablation. *According to Li J, Xu S. Past, present and future of ICP-MS instruments. Mod Sci Instrum 2011, 30(5).*

determined by IC, atomic absorption spectroscopy, ICP-AES, and ICP-MS. After vacuum thermal explosion of the fluid inclusions, the extracted gaseous-phase components are determined by QMS, GC, or GC-MS. In the following sections we introduce several common instruments and methods for gas and liquid component analysis of inclusions.

3.4.2.2.1 Inductively coupled plasma mass spectrometry

ICP-MS is currently recognized as the most authoritative elemental analysis technique. With the development of basic research and instrumentation, ICP-MS has greater advantages in isotope ratio analysis than any other analytical technique. With a high sensitivity (10^{-9}), low detection limit (up to parts per billion), and a simple mass spectrum, ICP-MS is suitable for the determination of Na, Mg, Mn, Zn, Cu, Pb, Sr, Ba, Rb, rare earth elements, and isotopes in the liquid phase of inclusions. Fig. 3.9 shows the main structure hardware of ICP-MS.

3.4.2.2.2 Ion chromatography method

As a branch of high-performance liquid chromatography, IC is the most common method for inorganic anion analysis. With the commercialization of IC in 1975, it has evolved from the initial analysis of common inorganic anions to various inorganic and organic ions. In traditional liquid-phase composition analysis of population colonies, the anions are described by single-channel IC. The components are measured without separating the inclusion solution after bursting at high temperature. The sensitivity can reach the order of parts per million to parts per billion (10^{-6}-10^{-9}), with simple operation, fast speed, small sample volume (100 μm), and low cost.

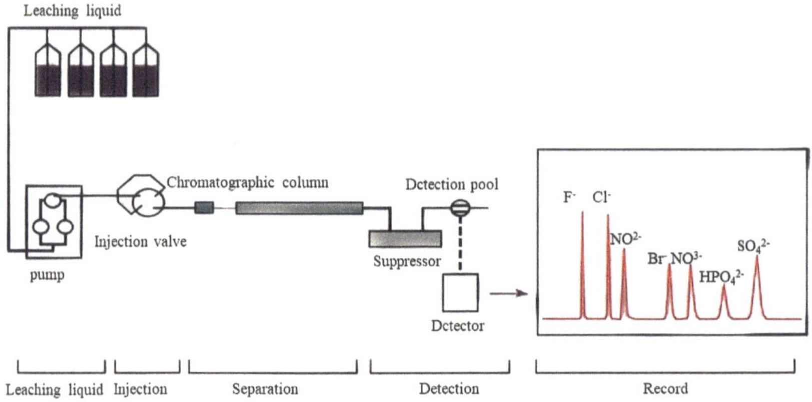

FIGURE 3.10 Basic flow of ion chromatography.

Various anions including F^-, Cl^-, Br^-, SO_4^{2-}, NO_3^-, and PO_4^{3-}, can be measured in one injection. In the 1980s and 1990s, IC was widely applied in the analysis of the anions in fluid inclusions with remarkable achievement of results.

IC consists of detection and analysis systems (Fig. 3.10). The main components are the conductivity detector, separation column (ion-exchange resin column with low exchange capacity), and suppression column (ion-exchange resin column with high exchange capacity). The accessories include a liquid storage tank, high-pressure pump, and injection valve. The leaching and test solutions are respectively injected into the instrument by the high-pressure pump and injection valve. An exchange reaction occurs after the hybrid leaching and test solutions are sent to the separation column. When the exchange reaches a dynamic equilibrium, the ions to be tested are adsorbed on the exchange resin. Dynamically flowing liquids separate the anions of different properties. The leaching liquid containing the ions to be tested enters into the conductivity pool through the suppression column. The obtained signal is amplified and sent to the recorder, where the anion is characterized and quantified based on the peak time and area.

Two-channel IC has been used to achieve simultaneous determination of anions and cations in liquid inclusion components. For example, Dan Yang et al. [11] used a two-channel IC to realize the simultaneous analysis of Li^+, Na^+, K^+, Ca^{2+}, F^-, Cl^-, and Br^- in the liquid components of fluid inclusions of quartz, calcite, fluorite, sphalerite, garnet, magnetite, and pyrite. Due to its lack of complexity, low cost, and low time consumption, IC has expanded the analysis range of mineral species, providing more direct and effective information for ore-forming fluid research.

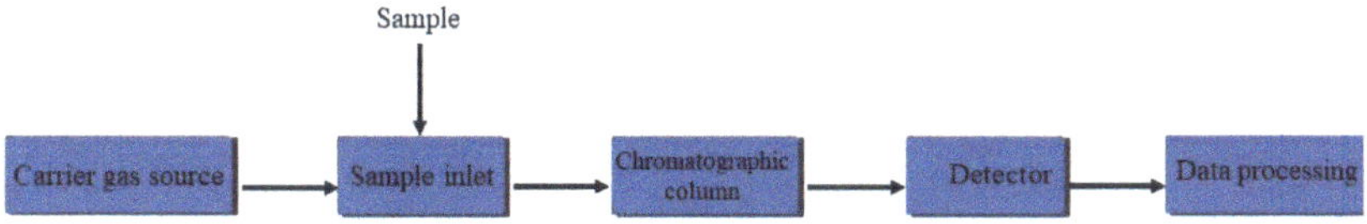

FIGURE 3.11 Basic flow pattern of gas chromatography.

3.4.2.2.3 Gas chromatography

GC is a column chromatography technique using an inert gas (N_2, He, Ar, H_2, etc.) as the mobile phase. The mixed gas is separated based on differences in physical and chemical properties, such as the boiling point, polarity, and adsorption of the substance. Fig. 3.11 shows a chromatographic system including the pure carrier gas source, sampling inlet (vaporizer for liquid sample), chromatographic column, and detector (for separation over time). As the components pass, the signal output value of the detector changes to respond to the components.

In geology, vacuum explosion is generally used to open fluid inclusions, and the gaseous-phase composition of the inclusions is analyzed by GC. Dan Yang et al. reported the application of a modified GC, GC-2010 (see Fig. 3.12). Two-dimensional GC with a double column and double detector in series was established to accurately determine H_2, O_2, N_2, and methane in fluid inclusions.

3.4.2.2.4 Quadrupole mass spectrometry

QMS is also one of the important methods for gaseous-phase composition analysis of fluid inclusions. The mass analyzer consists of four poles that are parallel to the axis and are equidistantly suspended. DC and RF voltages with equal amplitudes are supplied to the four poles.

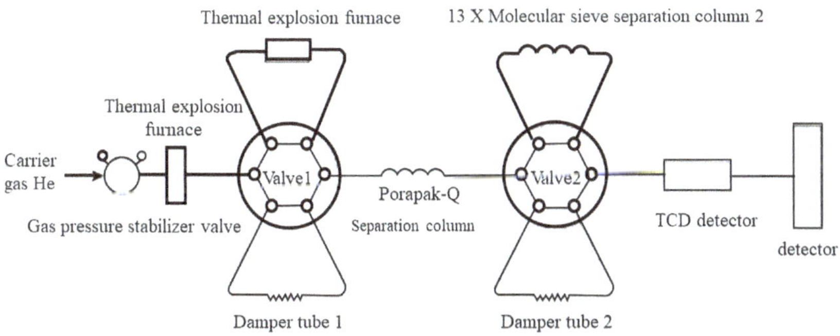

FIGURE 3.12 Diagram of the modified gas chromatograph GC-2010. *TCD*, thermal conductivity detector. *According to Yang D, Xu W, Cui Y. Determination of gaseous phase components in fluid inclusions by two-dimensional gas chromatography. Rocks Miner 2007;26(6):451–4.*

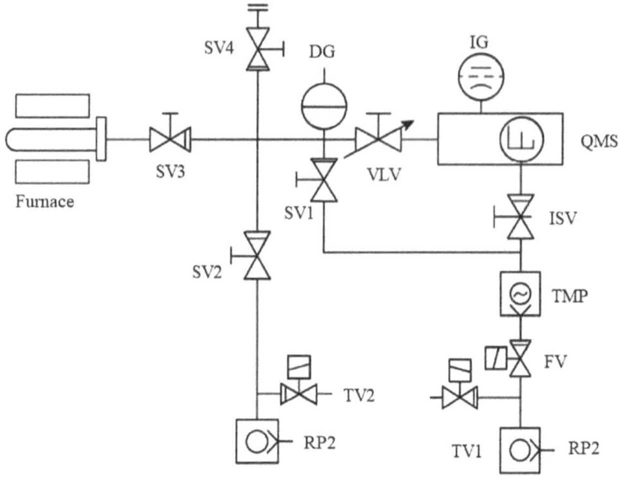

FIGURE 3.13 Diagram of quadrupole mass spectrometer measurement system. *SV1–SV4* and *ISV*, gas control valves; *VLV*, needle valve; *TMP*, molecular pump; *RP1* and *RP2*, mechanical pumps; *DG*, membrane manometer; *IG*, vacuum ionization gauge; *QSM*, quadrupole mass spectrometer; *TV*, transition valve; *FV*, furnace valve. *According to Zhu H, Wang L, Liu J. Determination of gaseous phase composition of fluid inclusions in different metallogenic stages by quadrupole mass spectrometry. Acta Petrol Sin 2003;19(2):314–8.*

The principle of QMS is that the different ionic mass-to-charge ratios lead to different trajectories of alternating electric fields generated by four parallel rods. Thereafter, the ions of the different masses are separated to obtain the mass spectrum of the molecules and atoms by the detector. By processing the mass spectrum obtained, the sample can be characterized and quantified. Heping Zhu applied the QMS RG202 of Japan Vacuum Technology to determine the gaseous-phase composition of fluid inclusions at different mineralization stages. Fig. 3.13 shows the QMS measurement system with specific operation as follows. First, 50 mg of quartz sample is put into a quartz glass tube. Then, the sample in the quartz glass tube is heated to 100°C by an electric furnace, and the SV2 valve is opened for evacuation. SV2 is closed so as to open SV1. When the vacuum degree in the tube reaches 5×10^{-6} Pa, the secondary inclusions and sample adsorption gas below 100°C are removed. SV1 is then closed to turn on the VLV valve for measurement. Meanwhile, the temperature of the electric furnace is raised to the set temperature at 1°C/5 s for instantaneous detection, thus obtaining a curve of total gas pressure and temperature, namely the burst-temperature curve. The burst-temperature curve indicates the different temperatures at which bursting occurs in the group inclusions. On this basis, the gaseous-phase components are measured within different temperature intervals.

3.4.3 Analysis of the chemical composition of individual inclusions

The compositions of individual inclusions can be determined by destructive (LA-ICP-MS, SEM/EDS, SIMS) and nondestructive (LRM, FTIR, SXRF, nuclear microprobe) methods. Table 3.3 shows the common analytical instruments and characteristics of individual fluid inclusions in minerals.

The common test instruments and methods based on nondestructive and destructive analysis of individual fluid inclusions are described below.

3.4.3.1 Destructive analysis of individual fluid inclusions

3.4.3.1.1 Laser ablation inductively coupled plasma mass spectrometry

LA-ICP-MS is the most popular analytical method for the liquid components of individual fluid inclusions. It is a destructive analytical method, and integrates laser ablation with ICP-MS. The specific principle is described as follows. First, a conical hole is made through to the inside of the inclusion to extract the liquid phase from the host crystal. After that, the multielement and isotope composition analysis is performed in ICP-MS to obtain the fluid composition information of the individual fluid inclusion. The sensitivity of ICP-MS for fluid inclusion detection reaches 10 μm.

Without the complexity of composition analysis and the uncertainty of data interpretation in block inclusions, LA-ICP-MS still has many problems that are difficult to solve in actual test operations. During laser hole drilling, the excessive temperature generated by the laser results in fissures in the host crystal from which the liquid-phase components of the inclusions leak out. In ICP-MS analysis, the presence of host crystal particles in the extracted fluid easily affects the results of the components analysis, causing large error. Günther et al. [14] discussed the complex multiphase analysis of inclusions by segmented ablation (see Fig. 3.14). Segmental ablation can effectively reduce erosion loss and host crystal interference, thus improving the method.

The upper matrix of the inclusion in the sample is first ablated with low laser power and a small hole from 4 to 10 μm (see Fig. 3.14A). The diameter of the hole is then increased to 20 μm for ablation until the inclusion is opened. For open inclusion, the internal gas- and few liquid-phase components enter the carrier system. Most liquid-phase components and mineral crystals remain in the inclusions (see Fig. 3.14B). Last, the whole inclusion is ablated with high laser power through a hole of 40 μm. The remaining components in the inclusion are all sent to the ICP as a laser aerosol (see Fig. 3.14C). The internal components of the fluid inclusions are obtained by segmented ablation to exclude the influence of the matrix on the inclusion components. Thus the loss of some components in individual inclusions can be reduced to improve the accuracy of the analysis.

TABLE 3.3 Common methods for the composition analysis of fluid inclusions.

Analysis object	Analytic method	Characteristics of analytical method	Sampling method
Block analysis	ICP-MS	For detection of rare earth elements in fluid inclusions	Crushing (or bursting) —extraction
	QMS	Analysis of gas components (mainly rare gases) in inclusions	
Analysis of individual inclusions	FTIR	To analyze inclusions in petroleum	Direct analysis of individual inclusion
	LA-ICP-MS	Has great advantages for the detection of low-content elements in fluid inclusions; however, it cannot eliminate the mutual interference between some atoms	
	SEM	Qualitative and semiquantitative determination of solids released in inclusions	
	PIXE or PIGE	According to different radiation-induced modes, it can be divided into PIXE and PIGE. Heavy elements can be detected by PIXE. PIGE detection is suitable for atomic nuclear energy elements undergoing p-r and p-a-r transformation.	
	SXRF	The error of volume estimation is reduced by analyzing massive inclusions close to the surface, such as anions, metal ions, and gases (Cl, Br, S, Ca, Mn, Pb, CO_2, etc.)	
	LRM	Analysis of components in inclusions with multiple phase states (e.g., minerals, CH_4, CO_2, etc.)	
	LMNGS	Quantitative analysis of rare gases and elements such as K, I, Te, Ca, and U in the inclusions	
	Ion probe	Analysis of C, H, N, F, and H_2O in fluids	

FTIR, Fourier transform infrared spectroscopy; *ICP-MS*, inductively coupled plasma mass spectrometry; *LA-ICP-MS*, laser ablation inductively coupled plasma mass spectrometry; *LMNGS*, laser microprobe noble-gas mass spectrometer; *LRM*, laser Raman spectroscopy; *PIGE*, particle-induced γ-ray emission; *PIXE*, particle-induced X-ray emission; *QMS*, quadrupole mass spectrometry; *SEM*, scanning electron microscopy; *SXRF*, synchrotron radiation X-ray fluorescence.
According to Lu H, Guo D. Progress and direction of fluid inclusion research. Geol Rev 2000;46(4):385–392.

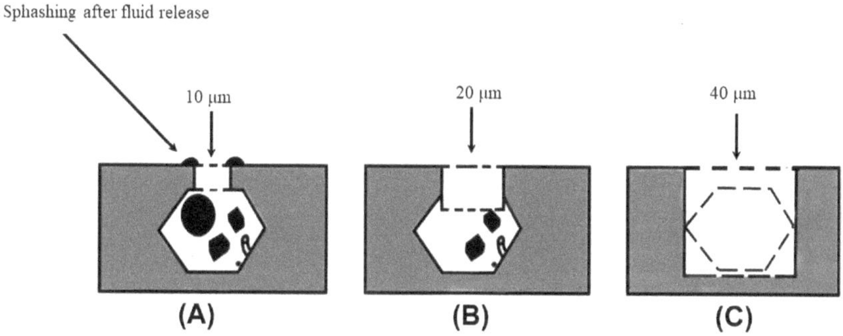

FIGURE 3.14 Segmented laser ablation of fluid inclusions. (A) 0.25-mm laser aperture (to produce 10-μm ablation holes); (B) 0.5-mm laser aperture; (C) 1-mm laser aperture. *According to Günther D, Audétat A, Frischknecht R, et al. Quantitative analysis of major, minor and trace elements in fluid inclusions using laser ablation–inductively coupled plasmamass spectrometry. J Anal Atomic Spectrom 1998;13(4):263–70.*

3.4.3.1.2 Scanning electron microscopy/energy-dispersive X-ray spectroscopy analysis

As the main instrument for research on daughter minerals in inclusions, SEM/EDS is used to analyze the morphology of opened inclusions as well as the minerals and the characteristics of solid components in inclusions. It has a good effect on fluid inclusion daughter minerals and melt inclusion components. This topic was partially introduced in the morphology analysis of inclusions in this chapter. Based on careful research of inclusions under an optical microscope, SEM/EDS is suitable for analysis of black opaque daughter minerals that cannot be identified under an optical microscope or that are less than 1–2 μm.

Sample preparation is as follows. The daughter minerals to be identified must be exposed on the surface of the test sample because the weak electron beam produced by SEM cannot pass through the cavity walls of fluid inclusions. There are two ways to expose the daughter minerals in the inclusions. First, the samples are observed under an optical microscope to find an inclusion sheet containing daughter minerals close to the surface. The test sheets are ground by hand under a microscope until the daughter minerals are clearly visible. However, this method is applicable only to small fluid inclusion samples. In other words, this method is not practically possible in inclusions with bulky daughter minerals, and when the constituent material is insoluble in water. Another method is to select a better mineral sample for crushing. Then, the fragments with high probability of daughter minerals are selected from the broken pieces. The fracture surfaces are fresh and flat, with the size of 10 × 10 mm and thickness of 3–5 mm. To prevent dehydration and contamination of the daughter minerals, the fragments should be placed in the drier immediately. Prior to the observation, the sample is coated with Au or carbon on the surface and then pasted on the sample holder for SEM.

The specific operation process is described as follows. First, put the sample under the SEM/EDS apparatus. Then slowly move it under low magnification of the microscope (500×−1000×) by rotating the x and y axes. The observation point is moved from left to right and from top to bottom to find small pits on the surface of the host crystal. Some small pits are likely to expose the inclusions. Once an inclusion is found, the magnification is increased by 2000×−5000× or more for further confirmation. After observation and photographing, the daughter minerals found in small pits (envelope voids) are analyzed by X-ray photoelectron spectroscopy to print the X-ray spectrum. The components and possible names of the daughter minerals are determined based on the daughter mineral morphology and X-ray characteristics.

3.4.3.1.3 Secondary ion mass spectrometry analysis

SIMS is mainly used for sample surface analysis. The primary ions hit the target surface to sputter the secondary ions. These ions are analyzed by mass spectrometry to obtain the surface information of the sample. SIMS can obtain information on isotopes, compound components, and molecular structure by analyzing all elements, including hydrogen. SIMS has high sensitivity and a low detection limit. For example, the detection limit of typical time-of-flight (TOF)-SIMS can reach parts per billion. TOF-SIMS also has other functions such as deep analysis and 3D reconstruction of sample surfaces, since it analyzes the internal components of the sample in layers. Therefore, it can be easily used to analyze the components of individual inclusions. The following section focuses on the application of TOF-SIMS in the analysis of individual fluid inclusions.

The working principle is shown in Fig. 3.15. TOF-SIMS bombards the mineral surface microregion (500−0.2 μm) by emitting kiloelectron volt-energy primary ions from the beam spot. The mineral surface atoms generate secondary ions by sputtering. The surface of the mineral is ablated by the continuous bombardment with primary ions, gradually forming a deep pit, which eventually opens the inclusion. The fluids can be diffused in the sample chamber with high vacuum. The chemical composition and structure of the inclusions are obtained by mass spectrometry. According to reports, the TOF-SIMS ablation rate is 10−20 μm/min, with an ablation depth of 150 μm.

There are several issues to be aware of in TOF-SIMS analysis:

(1) Positioning of inclusions: TOF-SIMS is analyzed under a reflected light. The positioning is a big problem for inclusions that can be seen only under transmitted light. Before the TOF-SIMS analysis, the inclusions are positioned and marked under a polarizing microscope. However, there are still errors and inconveniences.

(2) Sample conductivity: For nonmetallic minerals with poor conductivity, such as quartz, feldspar, and calcite, the mass spectrum obtained has a weak signal and large noise, which is unfavorable for the interpretation of a high-resolution mass spectrum. Therefore, sample surfaces with poor conductivity are first processed by surface charge function.

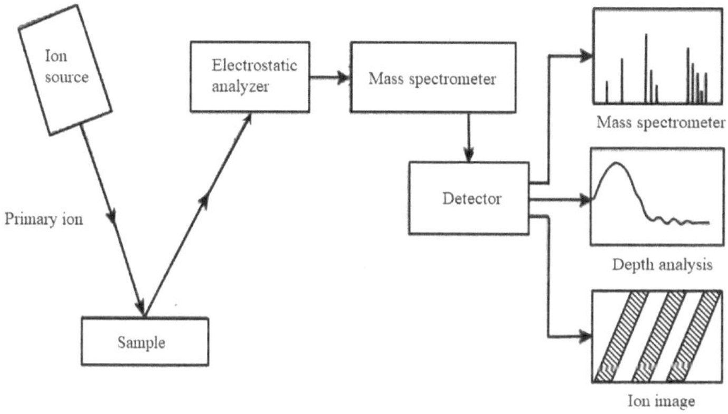

FIGURE 3.15 Diagram of time-of-flight secondary ion mass spectrometry operation.

(3) Identification of spectrum: With extremely high sensitivity and resolution, the TOF-SIMS can detect almost all substances present on the surface of the sample. In the mass spectrum, there are massive peaks including almost all mass numbers. Each peak should be analyzed in detail to obtain the ions present and their characteristics. In particular, some strong and weak peaks are generated due to experimental and environmental factors. The peaks with poor reproducibility need careful identification.

3.4.3.2 Nondestructive analysis of individual fluid inclusions

3.4.3.2.1 Laser Raman spectroscopy

As an emerging microregion analysis technique, LRM has high-precision, in situ, nondestructive, and fast characteristics. It has gradually become an important method for analysis in inclusion component research.

LRM can be used to analyze the gaseous-phase components in individual inclusions quickly and accurately by qualitative and semiquantitative methods of analyses. It is an important nondestructive analysis method for fluid inclusions. The working principle is the "Raman effect." When light passes through a substance, the partial scattered rays change in frequency. The spectrum of the light is called the Raman spectrum. According to the selection rule of spectrum, only those vibrations with changing molecular polarizability have Raman activity in the substance molecules, and the Raman spectrum is caused by these vibrations. As a combination of microprobes and ordinary Raman spectroscopy, microprobe Raman spectroscopy can analyze microscale microregions. The specific analysis steps are as follows. First, the sample is ground to a sheet with thickness of 100–300 μm according to the conventional method for producing inclusion sheets. Then the laser beam is used to align the gaseous-phase components in the inclusions determined by the microscope. The laser Raman data for each gas obtained by computer analysis can be measured by a standard gas to calculate the F value. The mole fraction of each gas is calculated based on the F value. Fig. 3.16 shows the structure of LRM.

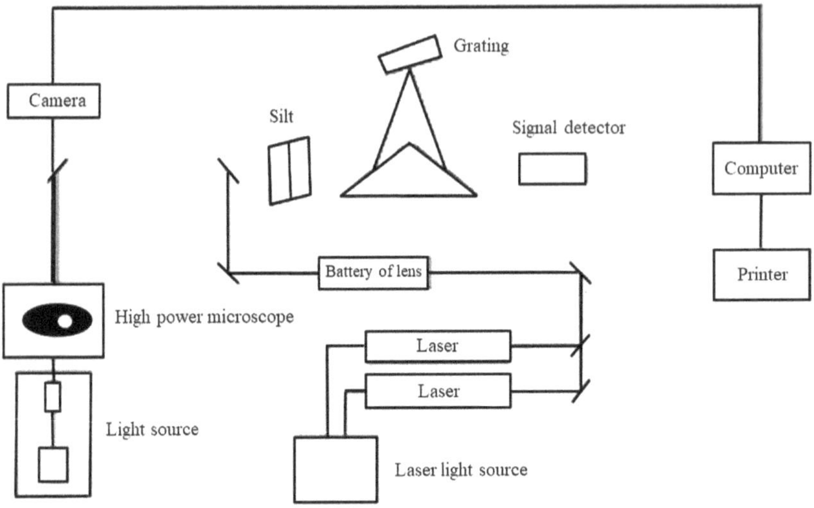

FIGURE 3.16 Structure of a laser Raman spectroscope. *According to Lu H, Fan H. Ni Pei fluid inclusion, Beijing: Science Press; 2004.*

LRM can also be used to measure the liquid-phase and daughter mineral components in the inclusions. Some scholars have also applied LRM to determine the salinity of fluid inclusions. For research on inclusions, LRM is a convenient and fast method and does not destroy the inclusions.

3.4.3.2.2 Fourier transform Infrared

FTIR directly provides molecular structure information on the individual fluid inclusions. Similar to LRM, it is also a nondestructive method of analysis for inclusions. The LRM and FTIR methods have the same characteristics; they include: (1) fundamental vibrations based on crystals and molecules; (2) fixed-point microregion analysis of solid, gas, and liquid inclusions; (3) quantitative analysis of the chemical properties of the substances in the inclusions; and (4) identification of organic compounds, daughter minerals, and solid components in the inclusions. Neither method can determine the composition of monoatomic (ion) substances or rare gases.

The working principle of FTIR is described as follows. FTIR consists of an infrared optical table, a computer, and a printer. Fig. 3.17 shows the structure of the FTIR. The main part, the infrared optical table, consists of an infrared light source, diaphragm, interferometer, sample chamber, detector, infrared mirrors, helium–neon laser, and control circuit and power supply.

FTIR is obtained from the Fourier transform of the data points acquired by the Michelson interferometer during the moving mirror scanning. Within an equal finite distance and position, the moving mirror collects the data points to form an interference pattern, thus obtaining an infrared spectrum by Fourier

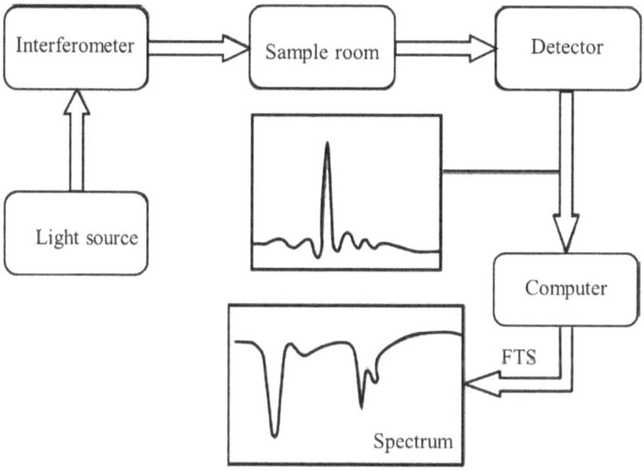

FIGURE 3.17 Basic optical path of Fourier transform infrared spectroscopy. *FTS*, Fourier transform spectroscope.

transform. Each data point consists of the data of x and y axes, which depend on the abscissa and ordinate units of the spectrum before the data acquisition. The abscissa unit can be expressed as a wave number or wavelength, of which the former is more popular. The ordinate is expressed as transmittance or absorbance in the transmission spectrum measured by the transmission method. The transmittance spectrum intuitively indicates the absorption of infrared light on the sample, though it is not proportional to the quality of the sample. In other words, the infrared spectrum cannot be quantitatively analyzed by the transmittance. In contrast, the value of the absorbance spectrum is proportional to the thickness and concentration of the sample within a certain range. Therefore, the absorbance is generally used to represent the infrared spectrum.

The samples for the test are prepared as follows. As a transmission technique, infrared microscopy uses emitted infrared beams to detect inclusions, atmosphere, carrier medium, and mineral matrix on the optical path. For this reason, sample preparation will be more rigorous and complex compared with its counterpart, LRM. Due to this experimental approach, the samples cannot be placed on infrared-absorbing media (such as standard slides). The sample sheets should be pasted with a clean two-sided polished inclusion sheet rather than typical adhesive (epoxy).

3.4.3.2.3 Synchrotron radiation X-ray fluorescence

Taking synchrotron radiation as the excitation source, SXRF illuminates different substances to emit different secondary X-rays. According to the X-rays of the substances, the intensities can be qualitatively and quantitatively determined. SXRF reaches a detection limit on the order of 10^{-9}, with high brightness, continuous energy spectrum, monochromatic adjustment, good accuracy, and strong polarization. Therefore, it has a better spatial resolution, analysis speed, and sensitivity than similar techniques in research on spatial distribution and content of trace elements. As a nondestructive detection technology, SXRF analyzes elements with long diameter of over 2 μm and atomic number over 13. Therefore, it is an ideal method for nondestructive quantitative analysis of individual fluid inclusions. The most critical limitation is the detection limit, including the maximum limit of sensitivity—X-ray absorption of incident fluorescent beam by the host crystal. The following methods are adopted to improve the analysis accuracy and reduce the detection limit:

(1) Fluid inclusions with smaller distance from the sheet surface are selected to increase the counting rate, reducing the detection limit of the test elements. The host crystal, instead of transition and heavier elements, easily absorbs the X-rays radiated by light elements. Therefore, SXRF detects transitional, heavier, and lighter elements with less depth in fluid inclusions that are deep beneath the mineral surface. Although it has simplicity, it can hardly solve the detection of fluid inclusions 20 μm below the mineral surface nor the absorption of the host crystal to the X-rays of the incident fluorescent beam.
(2) The absorption mode of the corrected X-ray is established to improve the accuracy and sensitivity of the analysis.

The preparation of the sample and the test are described as follows. The inclusions are ground to sheets with polished surfaces and thickness of 100–300 μm (in general, the thickness is 70–100 μm). After the sheet is removed from the slide, the adhesive on the sheet is washed with acetone or chemically pure ether. Then the sheet is attached to a film with no interfering elements, and fixed to a standard slide holder, thus preventing the breakage of the sheet. Due to the limited spatial resolution of the instrument, the representative inclusions are selected with sizes larger than 20 × 20 μm.

3.4.3.2.4 Microbeam proton–induced X/γ-ray analysis

Microbeam proton–induced X/γ-ray analysis belongs to SPM (scanning proton microprobe). In this technique, a high-energy proton microbeam interacts with the target substance to produce physical effects such as excitation of characteristic X-rays, scattering, transmission of protons, and generation of γ-rays by nuclear reactions. Therefore, SPM can be used to discuss the microscopic features of objects by nuclear effects, such as particle-induced X-ray emission (PIXE), Rutherford backscattering spectrometry, scanning transmission ion microscopy, elastic recoil detection analysis, and particle-induced γ-ray emission. PIXE is widely used in the analysis of elements with atomic number greater than 13. Due to the predictability of the proton trajectories, PIXE yield, X-ray absorption, high spatial resolution, high

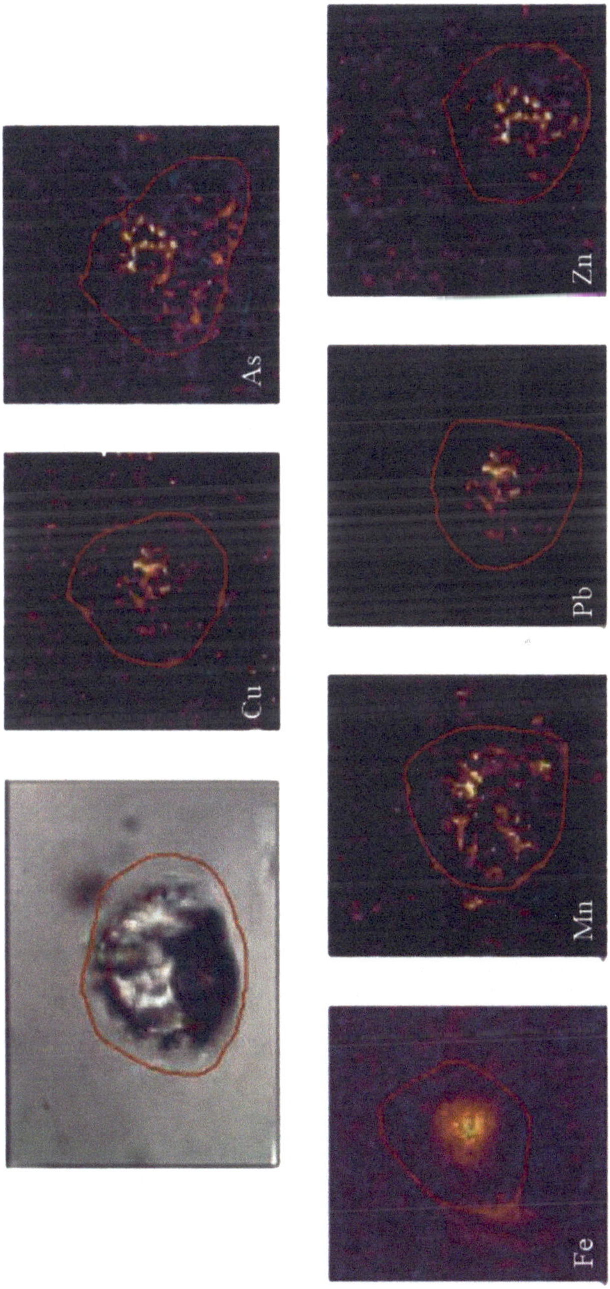

FIGURE 3.18 Optical micrograph of an individual inclusion and PIXE diagrams of individual multiphase inclusion elements (the first is an optical micrograph, and the rest are PIXE diagrams. *According to Fulignati P, Kamenetsky V, Marianelli P, et al. PIXE mapping on multiphase fluid inclusions in endoskarn xenoliths of AD 472 eruption of Vesuvius (Italy). Period Mineral 2013;82(2):291–7.*

sensitivity, and large reaction cross section, PIXE has become an ideal tool for elements with atomic number not smaller than 13, especially greater than 30. Like SEM, PIXE achieves micrometer-scale accuracy for simultaneous determination of multiple elements in fluid inclusions. However, its sensitivity is 2 orders of magnitude higher than that of SEM.

PIXE is affected by the beam overlapping, the shape, and the internal structure of inclusions. These factors are weakened by calculating a multilayer model of PIXE yield from fluid inclusions and analyzing the X-ray intensity of the PIXE spectrum. It is possible for PIXE to quantitatively analyze fluid inclusion components. For example, Anderson et al. (1989) used proton-induced X-ray and γ-ray radiation methods to obtain composition data on daughter minerals and heavy elements (such as Fe, Mn, Cu, Zn, Pb, and Br) in fluid inclusions.

Fulignati et al. also applied PIXE to analyze the multiphase fluid inclusions in nepheline and clinopyrite in a mine at Italy, to obtain the internal element composition of the inclusions (see Fig. 3.18).

The PIXE composition analysis shows that these inclusions are rich in Fe, Pb, Zn, and As$\pm$Cu$\pm$Mn. In Fig. 3.18, the brightness of the element is proportional to the content. The results show that PIXE is a highly effective method for analyzing the composition of materials in multiphase fluid inclusions.

References

[1] Zhang W, Chen Z. Fluid inclusion geology. Wuhan: China University of Geosciences Press; 1993.

[2] Lu H, Fan H. Ni Pei fluid inclusion. Beijing: Science Press; 2004.

[3] Lu H, Li B, Shen K. Inclusion geochemistry. Beijing: Geological Publishing House; 1990.

[4] Kouzmanov K, Bailly L, Ramboz C, et al. Morphology, origin and infrared microthermometry of fluid inclusions in pyrite from the Radka epithermal Cu deposit, Srednogorie zone, Bulgaria. Miner Depos 2002;37(6- 7):599–613.

[5] Shuey RT. Semiconducting ore minerals. Elsevier; 2012.

[6] Li N, Chen YJ, Ulrich T, et al. Fluid inclusions study of the Wunugetu Cu–Mo deposit, Inner Mongolia, China. Miner Depos 2012;47(5):467–82.

[7] Lambrecht G, Diamond LW. Morphological ripening of fluid inclusions and coupled zone-refining in quartz crystals revealed by cathodoluminescence imaging: implications for CL-petrography, fluid inclusions analysis and trace-element geothermometry. Geochem Cosmochim Acta 2014;141:381–406.

[8] Kyle JR, Ketcham RA. Application of high resolution X-ray computed tomography to mineral deposit origin, evaluation, and processing. Ore Geol Rev 2015;65:821–39.

[9] Kyle JR, Mote AS, Ketcham RA. High resolution X-ray computed tomography studies of Grasberg porphyry Cu-Au ores, Papua, Indonesia. Miner Depos 2008;43(5):519–32.

[10] Li J, Xu S. Past, present and future of ICP-MS instruments. Mod Sci Instrum 2011;30(5).

[11] Yang D, Xu W, Cui Y. Determination of gaseous phase components in fluid inclusions by two-dimensional gas chromatography. Rocks Miner 2007;26(6):451–4.

[12] Zhu H, Wang L, Liu J. Determination of gaseous phase composition of fluid inclusions in different metallogenic stages by quadrupole mass spectrometry. Acta Petrol Sin 2003;19(2):314–8.

[13] Lu H, Guo D. Progress and direction of fluid inclusion research. Geol Rev 2000;46(4):385–92.

[14] Günther D, Audétat A, Frischknecht R, et al. Quantitative analysis of major, minor and trace elements in fluid inclusions using laser ablation–inductively coupled plasmamass spectrometry. J Anal Atomic Spectrom 1998;13(4):263–70.

[15] Fulignati P, Kamenetsky V, Marianelli P, et al. PIXE mapping on multiphase fluid inclusions in endoskarn xenoliths of AD 472 eruption of Vesuvius (Italy). Period Mineral 2013;82(2):291–7.

[16] Campbell AR, Rye D, Petersen U. A hydrogen and oxygen isotope study of the San Cristobal Mine, Peru; Implications of the role of water to rock ratio for the genesis of wolframite deposits[J]. Economic Geology 1984;79(8):1818–32.

[17] Hall DL, Sterner SM, Bodnar RJ. Freezing point depression of NaCl-KCl-H_2O solutions[J]. Economic Geology 1988;83(1):197–202.

Further reading

[1] Bodnar RJ. A method of calculating fluid inclusions volumes based on vapor bubble diameters and PVTX properties of inclusions fluids. Econ Geol 1983;78(3):535–42.

[2] Sun Y. Understandings on research and application of petroleum inclusions. Bull Miner Petrol Geochem 2006;1.

[3] Ge X, Su W, Zhu L. Effect of infrared light intensity on salinity of fluid inclusions in opaque minerals: taking stibnite as an example. Acta Mineral Sin 2011;31(3):366–71.

[4] Moritz R. Fluid salinities obtained by infrared microthermometry of opaque minerals: implications for ore deposit modeling—a note of caution. J Geochem Explor 2006;89(1):284–7.

[5] Lüders V, Ziemann M. Possibilities and limits of infrared light microthermometry applied to studies of pyrite-hosted fluid inclusions. Chem Geol 1999;154(1):169–78.

[6] Lüders V, Reutel C. Possibilities and limits of infrared microscopy applied to studies of fluid inclusions in sulfides and other opaque minerals. In: Pan-American Conference on Research on Fluid Inclusions (PACROFI) VI [J]. Madison, Wisconsin; May 1996. p. 78–80.

[7] Lindaas SE, Kulis J, Campbell AR. Near-infrared observation and microthermometry of pyrite-hosted fluid inclusions. Econ Geol 2002;97(3):603–18.

[8] Rosière CA, Rios FJ. The origin of hematite in high-grade Fe ores based on infrared microscopy and fluid inclusions studies: the example of the Conceição mine, Quadrilátero Ferrífero, Brazil. Econ Geol 2004;99(3):611–24.

[9] Gu Du, Wang K, Jing R. Rock mineral identification technology and its application in modern large-scale instruments such as infrared spectroscopy/scanning electron microscopy. Rock Miner Anal 2014;33(5):625–33.

[10] Lin C, Miller J. Network analysis of filter cake pore structure by high resolution X-ray microtomography. Chem Eng J 2000;77(1):79–86.

[11] Lin C, Miller J. Cone beam X-ray microtomography-a new facility for three-dimensional analysis of multiphase materials. Miner Metall Process 2002;19(2):65–71.

[12] Zhang X. Application conditions of determination of mineral inclusion salinity by freezing method. Geol Prospect 1993;9:008.

[13] Lu X, Yao S, He M. Raman spectroscopic determination of salinity of ore-forming fluid inclusions. Geosci Front 2001;8(4):429–33.

[14] Li B, Yang H. Application of inductively coupled plasma mass spectrometry (ICP-MS) in geoscience research. Geosci Front 2003;10(2):367–78.

[15] Yang D, Xu W. Simultaneous determination of liquid anion and cation in various mineral fluid inclusions. Acta Petrol Mineral 2014;33(3):591–6.

[16] Wang L. Analysis of composition of fluid inclusions. Geol Rev 1998;44(5):496–501.

[17] Diamond LW, Marshall DD, Jackman JA, et al. Elemental analysis of individual fluid inclusions in minerals by Secondary Ion Mass Spectrometry (SIMS): application to cation ratios of fluid inclusions in an Archaean mesothermal Au-quartz vein. Geochem Cosmochim Acta 1990;54(3):545–52.

[18] Li R, Zhou S. Application of single organic shell test and TOF-SIMS technology in minerals. Acta Mineral Sin 2000;20(2):172–6.

[19] Sun Q, Zeng Y. Progress in nondestructive analysis of single fluid inclusions. Adv Earth Sci 2000;15(6):673–8.

[20] Anderson AJ, Clark AH, Ma XP, et al. Proton-induced X-ray and gamma-ray emission analysis of unopened fluid inclusions. Econ Geol 1989;84(4):924–39.

Chapter 4

Internal composition of mineral fluid inclusions

Chapter outline

4.1 Gaseous-phase composition of inclusions 70
4.2 Liquid-phase components of inclusions 74
4.3 Solid-phase composition of inclusions 78
4.4 Metal components in fluid inclusions 80
4.4.1 Analysis and calculation of heavy metal elements in ore-forming fluids 80
4.4.1.1 Sample selection 80
4.4.1.2 Sample preparation 82
4.4.1.3 Sample test 82
4.4.1.4 Dilution multiples and concentration conversion 82
4.4.2 Metal components in fluid inclusions in Cu–Au mineral 83
4.4.3 Metal components in fluid inclusions of Pb–Zn mineral 85
4.4.4 Metal components of group inclusions in quartz and pyrite 88
References 95
Further reading 96

Mineral fluid inclusions have very complicated internal compositions, which are closely related to the type of deposit, the ore-forming environment, and the geological evolution at the late stage of inclusion formation. In general, the internal substances of fluid inclusions are divided into gaseous, liquid, and solid phases according to their existing form. The gaseous-phase components in the inclusions are of little significance to the mineral processing discipline. However, the composition of the liquid and solid phases is of great significance to the mineral processing discipline. The fluid inclusions in mineral ore are opened during grinding due to physical damage by external forces, which causes the release of internal materials, especially the liquid components, into the ore pulp solution, which inevitably causes changes in the chemical environment of the ore pulp solution and the surface properties of the mineral, thus affecting the flotation of the mineral.

Existing research data indicate that the most common gaseous-phase components in inclusions are CO_2, F, Cl, CH_4, CO, H_2, H_2O, O_2, N_2, and some gaseous hydrocarbons. The solid-phase materials in mineral inclusions mainly

Fluid Inclusion Effect in Flotation of Sulfide Minerals. https://doi.org/10.1016/B978-0-12-819845-2.00004-1

comprise the daughter minerals in the inclusions. The common solids in fluid inclusions are halite and potassium salts, and small amounts of sulfides, sulfates, carbonates, phosphates, silicates, borates, and metal oxides are also present. The liquid-phase components of inclusions are complicated. The main components are the same as those of the ore-forming fluids during the minerals' formation. Generally, the anions are mainly Cl^-, F^-, and SO_4^{2-}; the cations are alkali metal ions such as Na^+, K^+, Ca^{2+}, and Mg^{2+} and some heavy metals. According to the fluid chemistry of mineral formation, the fluid inclusion component of nonferrous metal sulfide minerals must contain the same-named heavy metal ion—one of the host crystal components. For example, fluid inclusions in chalcopyrite mineral must contain a Cu component; fluid inclusions in sphalerite must contain Zn. Due to the concomitant relationship between the minerals in the ore-forming process and the chemical diversity of the ore-forming fluids, partial ore-forming fluids are captured between the main and the gangue minerals. The specific situation depends on the nature of the deposit.

So far, the research on fluid inclusions in geochemistry has focused on inclusion morphology, freezing point temperature, sodium chloride content, fluid properties, and chemical composition. The chemical composition analysis focuses on the alkali metal rather than heavy metal. Geochemical scholars believe that there must be a homonymic metal ion in mineral inclusions, and heavy metal research has made only limited contributions to the information on mineralization. The following sections discuss mineral fluid inclusion with respect to components such as gases, alkali and alkali earth metals, and heavy metals present within inclusions.

4.1 Gaseous-phase composition of inclusions

The mineral inclusions in various deposits have complex gaseous-phase compositions, which are closely related to the type of deposit and the ore-forming environment. Research indicates that the most common gaseous-phase component in inclusions is CO_2; others are F, Cl, CH_4, CO, H_2, H_2O, O_2, N_2, and some gaseous hydrocarbons.

In China's carbonate rocks, large-scale Mississippi Valley-type and Huayuan Pb–Zn deposits are taken as examples. Geologists found that there are fluid inclusions in the sphalerite, calcite, barite, and fluorite of this deposit. The inclusions are long circles with diameters of about 5–15 μm and with irregular distribution. The main part is the single-phase liquid inclusion, next is the gas–liquid two-phase inclusion, with the gas–liquid ratio of 1%–3%. The gaseous-phase components in the inclusion were determined by thermal explosion gas chromatography. The gaseous hydrocarbons (C_1–C_4 alkanes and olefins) were analyzed by vacuum milling–gas chromatography (see Tables 4.1 and 4.2).

According to this analysis, there are many gases in the mineral inclusions of the research region, including CO_2 (1.1–12.0 mol%) and gaseous hydrocarbon (0.227–3.17 mol%). The Duobuza Cu deposit in the Ali area of Tibet is the first large Cu deposit discovered on the Bangong Lake–Nujiang metallogenic belt.

TABLE 4.1 Characteristics of gaseous-phase composition (mol%) of the Huayuan Pb–Zn deposit.

Mineral	Sample characteristics	Sample number	H_2O (%)	CO_2 (%)	Short-chain hydrocarbons			H_2 (%)	CO (%)
					Alkane		Olefins		
					Methane	Ethylene propylene	Ethylene propylene		
Sphalerite	Light yellow	7	91.53	6.38	0.93	0.12	0.28	0.21	0.57
	Brown	3	94.79	4.30	0.65	0.04	0.09	0.04	0.11
	Ore in limestone	6	93.92	4.60	0.86	0.04	0.11	0.08	0.39
	Ore of argillaceous limestone	4	90.64	7.18	0.82	0.17	0.39	0.28	0.44
	North mining area	6	91.62	6.45	1.00	0.11	0.23	0.16	0.44
	South mining area	4	93.84	4.71	0.60	0.07	0.21	0.15	0.43
	Average	10	92.51	5.75	0.84	0.09	0.22	0.13	0.14
Gangue	North mining area	3	94.63	3.69	1.30	0.04	0.08	0.13	0.14
	South mining area	4	94.64	4.11	0.58	0.05	0.08	0.43	0.12
	Average	7	94.63	3.79	0.89	0.04	0.08	0.30	0.13
Global average		17	93.38	5	0.86	0.07	0.16	0.22	0.31

According to Liu W, Zheng R. Research on gaseous phase composition of inclusions in Huayu Pb-Zn deposit—organic mineralization of MVT deposits (II). Acta Sedimentol Sin 1999;17(4):508–614

TABLE 4.2 Gaseous-phase composition (mol%) of fluid inclusions from the Huayuan Pb–Zn deposit.

Mining area	Sample characteristics	Chlorine phase component (mol%)				Organic gaseous phase component (mol%)		
		CO_2	H_2O	H_2	CO	CH_4	$C_{2-4}H_{6-20}$	$C_{2-4}H_{4-6}$
Naizibao	Brown sphalerite in limestone	4.51	95.20	0.02	/	0.24	0.01	0.02
	Light sphalerite in argillaceous limestone	6.97	89.61	0.25	1.23	0.95	0.34	0.65
	Light sphalerite in dolomitized argillaceous limestone	12.99	85.22	0.45	/	0.80	0.15	0.39
	Light sphalerite in limestone	5.62	93.06	0.10	1.04	0.14	0.02	0.03
	Calcite in limestone	2.15	95.29	0.09	0.07	2.27	0.05	0.08
	Barite in limestone	1.14	98.66	0.03		0.14	0.01	0.03
Banpo	Brown sphalerite in limestone	1.28	96.90	0.05	0.14	1.03	0.02	0.05
	Light sphalerite in limestone	6.81	89.72	0.10	0.20	2.86	0.10	0.21
	Calcite in limestone	7.79	89.93	0.28	0.34	1.49	0.06	0.13

Yutang	Brown sphalerite in limestone	6.56	92.26	0.04	0.20	0.68	0.06	0.20
	Light sphalerite in limestone	3.50	95.37	0.15	0.56	0.22	0.05	0.15
	Light sphalerite in argillaceous limestone	6.49	91.34	0.26	0.29	1.12	0.13	0.36
	Light sphalerite in dolomitized argillaceous limestone	2.27	96.37	0.14	0.65	0.39	0.04	0.14
	Calcite in limestone	2.20	96.66	0.12	0.03	0.88	0.08	0.02
	Calcite in limestone	6.84	92.06	0.61	0.13	0.26	0.03	0.08
	Calcite in argillaceous limestone	3.60	86.90	0.48	0.49	8.22	0.48	0.04
	Calcite in limestone	2.61	95.40	0.71	0.33	0.73	0.07	0.16
	Dolomite in limestone	4.77	94.43	0.27	/	0.46	0.01	0.05
Average		5.00	93.38	0.22	0.31	0.86	0.07	0.16
Standard deviation (δ)		3.00	3.34	0.20	0.36	0.75	0.08	0.16

According to Liu W, Zheng R. Research on gaseous phase composition of inclusions in Huayu Pb-Zn deposit—organic mineralization of MVT deposits (II). Acta Sedimentol Sin 1999;17(4):508–614.

Table 4.3 shows the gaseous-phase compositions of the fluid group inclusions of eight quartz samples in the mining area. Among the gaseous-phase composition of the eight mineral fluid inclusions in the three metallogenic stages, H_2O is the main component, followed by CO_2. There are also small amounts of CH_4, C_2H_6, N_2, and Ar and a trace of H_2S. The H_2O and CO in the inclusions show a gradual decrease and increase with the development of mineralization. In the quartz inclusions, the average contents of H_2O and CO_2 are 95.96% and 2.81%, respectively. There are small amounts of CH_4, C_2H_6, H_2S, N_2, and Ar, but the H_2O and CO_2 dominate the gaseous-phase composition of the quartz inclusions at various stages of mineralization, indicating an oxidizing environment.

4.2 Liquid-phase components of inclusions

The liquid-phase components of inclusions are complicated. As the key to mineral processing research, they are closely related to the nature of the deposit. As mentioned before, the main components of fluid inclusions are the same as those of the ore-forming fluids during the formation of the minerals. Generally, the anions are mainly Cl^-, F^-, and SO_4^{2-}; the cations are the alkali metal ions such as Na^+, K^+, Ca^{2+}, and Mg^{2+} and some heavy metals. In geology, the focus is on the determination of the alkali and alkaline earth metal ions and anions. There are few studies on heavy metal components. Table 4.4 shows the liquid-phase components of the fluid inclusions from the Duobuza Cu deposit.

The Duobuza Cu deposit has the following main characteristics. First, there are large amounts of Cl^-, SO_4^{2-}, Na^+, and K^+; the solutions also contain traces of Ca^{2+}, Mg^{2+}, and F^- that were not detected. Second, the solutions have high concentrations of Na^+ and Cl^-. Hence, there will be a large amount of NaCl in the fluid. The hydrochemical type of the Duobuza copper deposit fluids is $Cl^-{-}SO_4^{2-}{-}K^+{-}Na^+{-}Ca^{2+}$. The mineralization fluid is rich in Na, K, and Ca.

The Xihuashan tungsten deposit in Jiangxi is a large vein−type tungsten deposit with massive quartz fluid inclusions. The inclusions are large in volume and widely distributed. The main components of the inclusions are the gas and liquid phases, with particularly large volume (10−20 μm). Sometimes the volume is greater than 20 μm (not greater than 70 μm). The gas−liquid phase ratios are generally between 5%-15%. Few reach up to one-half. Table 4.5 shows the liquid-phase composition of the inclusions.

The liquid-phase analysis shows that the order of abundance of these anions is $Cl^- > SO_4^{2-} > F^-$. The Cl^- and SO_4^{2-} contents are more abundant than F^- in the sample. Of the cations, $Na^+ > K^+$, while the others are present in trace amounts. Therefore, the Xihuashan tungsten deposit is low in salinity, with the ion type of $Na^+{-}K^+{-}Cl^-{-}SO_4^{2-}$. The ratio of F^- to Cl^- is the smallest in quartz gangue and largest in tungsten mineral. This indicates that the concentration of tungsten is inversely proportional to that of fluorine in liquid phase. The Na^+/K^+ ratio is larger than 1, which indicates that there is much Na^+ in the solution. Most fluids with $Na^+/K^+ > 1$ are caused by magma.

The ions of alkali metals and alkaline earth metals and anions such as fluoride and chloride in the inclusions have less significant effect on mineral

TABLE 4.3 Results of the gaseous-phase composition analysis of fluid inclusions from the Duobuza Cu deposit.

			Gaseous phase composition (%)							
Sample number	Mineral	Mineralization stage	H_2O	N_2	Ar	O_2	CO_2	CH_4	C_2H_6	H_2S
DBZ060	Quartz	I	98.050	0.118	0.037	-	1.660	0.086	0.038	0.010
DBZ014	Quartz	II	97.070	0.160	0.052	0.137	2.255	0.127	0.075	0.124
DBZ044			98.020	0.145	0.045	0.006	1.598	0.106	0.053	0.028
DBZ045			97.430	0.154	0.045	0.041	2.036	0.122	0.113	0.059
DBZ059			97.750	0.106	0.033	0.068	1.839	0.084	0.049	0.071
DBZ025	Quartz and ggypsum	III	91.670	0.147	0.052	2.731	3.456	0.152	0.072	1.720
DBZ031			91.730	0.434	0.143	0.117	6.806	0.368	0.198	0.203
DBZ041			97.330	0.139	0.049	0.123	2.060	0.111	0.071	0.118

The results for Ar are for reference only.
According to He Y, Wen C, Liu X. Analysis of liquid phase composition of inclusions in multi-non-copper deposits. Metal Mine 2013;(3):108–110.

TABLE 4.4 Results of gaseous-phase composition analysis of fluid inclusions in the Duobuza Cu deposit ($\times 10^{-6}$).

Sample number	Mineralization stage	Anion content		Cation content				Hydrochemical type
		Cl^-	SO_4^{2-}	Na^+	K^+	Mg^{2+}	Ca^{2+}	
DBZ060	I	30.00	31.83	21.36	4.80	0.58	2.50	Cl^-–SO_4^{2-}–K^+–Ca^{2+}–Na^+
DBZ014	II	16.35	113.04	18.15	5.58	0.69	2.25	Cl^-–SO_4^{2-}–K^+–Ca^{2+}–Na^+
DBZ044	II	24.54	65.07	23.55	5.67	0.75	2.25	Cl^-–SO_4^{2-}–K^+–Ca^{2+}–Na^+
DBZ045	II	26.58	73.44	21.66	4.53	0.58	2.00	Cl^-–SO_4^{2-}–K^+–Ca^{2+}–Na^+
DBZ059	II	19.08	119.94	25.47	10.80	1.27	2.90	Cl^-–SO_4^{2-}–K^+–Ca^{2+}–Mg^{2+}–Na^+
DBZ025	III	6.81	327.87	10.53	19.20	0.35	2.50	Cl^-–SO_4^{2-}–K^+–Ca^{2+}–Na^+
DBZ031	III	17.73	114.21	23.40	10.80	1.13	2.35	Cl^-–SO_4^{2-}–K^+–Ca^{2+}–Mg^{2+}–Na^+
DBZ041	III	24.90	99.21	28.11	7.53	0.95	3.69	Cl^-–SO_4^{2-}–K^+–Ca^{2+}–Na^+

According to He Y, Wen C, Liu X. Analysis of liquid phase composition of inclusions in multi-non-copper deposits. Metal Mine 2013;(3):108–110.

TABLE 4.5 Liquid-phase composition of fluid inclusions of the Xihuashan tungsten deposit ($\times 10^{-6}$).

Sample nnumber	Rock name	Mineral		Cl^-	SO_4^{2-}	Na^+	K^+	Mg^{2+}	Ca^{2+}	Na^+/K^+	F^-/Cl^-
431-1-2	Quartz vein tungsten deposit	Quartz	1.32	4.44	17.8	7.41	1.98	<0.05	1.02	3.74	0.30
431-3-2	Greisen	Quartz	0.048	7.02	2.19	8.22	0.60	<0.05	<0.05	13.70	0.01
230-1-1	Quartz vein tungsten deposit	Quartz	0.54	12.7	20.7	7.26	1.47	<0.05	<0.05	4.94	0.04
1167-3	Quartz vein	Quartz	0.075	5.49	2.52	7.8	0.48	<0.05	<0.05	16.25	0.01
1167-6	Greisen	Quartz	0.135	8.64	1.98	9.96	0.48	<0.05	<0.05	20.75	0.02
1167-7	Quartz vein tungsten deposit	Quartz	0.129	3.78	1.47	5.76	0.75	<0.05	<0.05	7.68	0.03
1167-10	Quartz vein tungsten deposit	Quartz	0.153	5.79	0.60	6.87	0.69	<0.05	0.30	9.96	0.03
1167-13	Quartz vein tungsten deposit	Quartz	0.084	4.11	1.80	6.06	0.87	<0.05	<0.05	6.97	0.02
1167-15	Quartz vein tungsten deposit	Quartz	0.251	5.4	17.9	8.85	3.18	<0.05	<0.05	2.78	0.05
1124-1	Quartz vein tungsten deposit	Quartz	0.138	3.96	1.02	5.64	0.90	<0.05	<0.05	6.27	0.03
1124-3	Quartz vein tungsten deposit	Quartz	0.162	3.36	1.41	5.19	1.05	<0.05	<0.05	4.94	0.05
1185-16	Quartz vein	Quartz	0.084	6.54	0.93	8.67	1.14	<0.05	<0.05	7.61	0.01
1223-2	Quartz vein tungsten deposit	Quartz	0.129	1.92	1.38	3.96	1.38	<0.05	<0.05	2.87	0.07
1264-5	Quartz vein	Quartz	0.132	20.9	1.86	9.46	1.86	<0.05	<0.05	4.55	0.01

Test unit: Inclusion Laboratory, Institute of Geology and Geophysics, Chinese Academy of Sciences. Tester: Heping Zhu.
According to Xu T, Li Z. Fluid inclusion characteristics and source of ore-forming fluids in Xihuashan tungsten deposit, Jiangxi Province. Resour Surv Environ 2013;34(2): 95–101.

processing flotation than copper and lead ions. However, these components can be used as strong evidence of the release of inclusion components during the grinding process. Meanwhile, researchers have shown that the presence of calcium and magnesium ions has a significant impact on flotation; it was also shown by research that a large amount of collector (fatty acid) is consumed, especially in oxidized mineral flotation.

4.3 Solid-phase composition of inclusions

The solid-phase material in mineral inclusions mainly comprises the daughter minerals in the inclusions. The solids are directly generated from the fluid or melt in a relatively closed system. With a large growth space in the solution, the daughter mineral is generally a separate crystal. In the melt, the daughter minerals flow out of the walls of the inner chamber, and further evolve acording to the phase rule Gibbs. Research has shown that daughter minerals are mainly halides, some sulfides, salts, and metal oxides. The melt fluid inclusions are mainly composed of silicates and minerals. With the continuous development of high-end analytical techniques such as scanning electron microscopy/energy-dispersive X-ray spectroscopy (SEM/EDS) and electron probe microanalysis (EPMA), the research into daughter minerals in inclusions of hydrothermal deposits has made significant progress in recent years. Chalcopyrite daughter mineral is the most popular metal daughter mineral for scholars. Most chalcopyrite daughter minerals exist in porphyry Cu deposits, and some are found in sphalerite, chalcopyrite, and pyrite.

Yuling Xie et al. used SEM/EDS to discuss the fluid inclusions of the Taibai gold deposit of Shaanxi, obtaining SEM images of the daughter minerals in titanium-bearing dolomite fluid inclusions (see Fig. 4.1).

Quartz samples containing daughter minerals are selected to obtain major daughter minerals through research. For example, some single crystal minerals with cubic or column shapes have high peaks of Ca, Mg, and Fe ((Mg + Fe)/Ca = 0.833–0.729; Fe/Mg = 1.06–1.27) (see Table 4.6). Therefore, it must

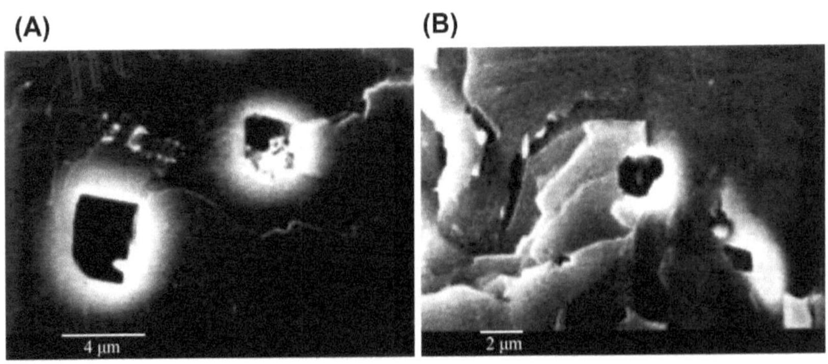

FIGURE 4.1 Scanning electron microscopy images of the daughter minerals in titanium-bearing dolomite fluid inclusions (according to Xie et al., 2002). (A) Single crystal of Fe-bearing dolomite daughter mineral in titanium-bearing dolomite. (B) Pyrite daughter mineral in titanium-bearing dolomite.

TABLE 4.6 Scanning electron microscopy analysis of daughter minerals in Fe-containing dolomite fluid inclusions in Taibai Au deposit.

	T440-2		T210-1[a]		T210-2①		T210-3		T552-2①	
Sample number	%	Atom%	%	Atom%	%	Atom%	%	Atom%	Wt%	Atom%
Fe	30.851	30.259	46.14	32.96	/	/	23.129	15.390	23.59	16.06
S	19.872	33.953	53.86	67.04	/	/	0.158	0.183	/	/
Na	/	/	/	/	35.05	45	0.00	0.000	/	/
Cl	/	/	/	/	64.95	55	0.903	0.947	/	/
Si	/	/	/	/	/	/	0.352	0.466	/	/
Ca	/	/	/	/	/	/	53.785	49.871	57.84	54.89
Mg	/	/	/	/	/	/	21.681	33.143	18.57	29.05
Zn	0.550	0.461	/	/	/	/	/	/	/	/
As	47.882	35.009	/	/	/	/	/	/	/	/
Te	0.557	0.239	/	/	/	/	/	/	/	/
Pb	0.296	0.078	/	/	/	/	/	/	/	/
Total	100.008	100.00	100.00	100.00	100.00	100.00	100.008	100.00	100.00	100.00
Host crystal	Quartz		Fe-containing dolomite		Fe-containing dolomite		Fe-containing dolomite		Fe-containing dolomite	
Daughter mineral	Arsenopyrite		Pyrite		Halite		Fe-containing dolomite		Fe-containing dolomite	

Scanning electron microscopy/energy-dispersive X-ray spectroscopy instruments: SEM S-250MK3 and EDS AN10000 with resolution of 50A. /, untested.
[a] *Raw data were extracted with impurity removal to calculate the percentage.*
According to Xie Y, Li S. Discovery of minerals such as pyrite and iron dolomite in fluid inclusions of Taibai Au deposit and its genetic significance. Miner Deposits 2000;19(1):54–60.

be Fe-bearing dolomite. Second, daughter minerals with polymorphic structure and complex morphology are analyzed by energy spectrum as Fe-bearing dolomite and halite. Finally, some daughter minerals are single crystals; these include pyrite, toxic sand daughter mineral, and Fe-bearing dolomite with certain crystalline forms.

When opaque daughter minerals on the surface of mineral inclusions are observed to be light yellow, they are metallic minerals. The daughter minerals are carefully polished to expose the surface for electron probe analysis. In the electron probe analysis of smaller daughter minerals, the surrounding host crystals are removed to avoid their influence on the experimental results, and it is found that all opaque daughter minerals on the surface of fluid inclusions are pyrite (see Table 4.7). In Table 4.6, SEM/EDS analysis demonstrates the presence of pyrite daughter minerals in the form of incomplete crystals.

4.4 Metal components in fluid inclusions

4.4.1 Analysis and calculation of heavy metal elements in ore-forming fluids

In geology, the quantitative analysis and content calculation of heavy metal elements in ore-forming fluids are of great significance to understanding the mechanism of formation and the process of hydrothermal mineral deposit. This analysis is equally important in the mineral processing field; it is used to estimate the heavy metal content in the liquid component. The components of the original ore-forming fluids are obtained by extracting the representative component liquid of the primary fluid inclusions. The solution obtained is usually diluted thousands of times; therefore it is necessary to restore the concentration of the solution for the analysis. In the following sections, we introduce the extraction of the liquid component of inclusions, the analysis of component concentrations, and metal content estimation of ore-forming fluids according to the measured concentration.

4.4.1.1 Sample selection

Proper selection of the samples is the first step in the analysis and calculation of metal elements in ore-forming fluids. The basic principles of the sample selection are described as follows: (1) Hydrothermal metal deposits with scheduled mineralization stages must be selected. (2) The main ore-forming minerals are selected to analyze the ore-forming process of the deposit. (3) Samples containing primary fluid inclusions should be selected. This is because when the ore-forming fluids meet the requirements it saves the process and improves the accuracy of the experiment.

TABLE 4.7 Daughter mineral electron probe analysis of Fe-containing dolomite fluid inclusions of Taibai Au deposit.

Sample number		Fe	S	Co	Ni	Zn	As	Te	Au	Ag	Pt	Total	Daughter mineral
T590[a]	%	45.73	53.40	0.05	0.00	0.16	0.14	0.19	0.13	0.06	0.13	99.99	Pyrite
	Atom%	32.84	66.81	0.04	0.00	0.10	0.08	0.06	0.03	0.02	0.03	100.01	
F15[a]	%	45.88	52.35	0.20	0.62	0.27	0.46	0.12	0.00	0.10	0.00	00.00	Pyrite
	Atom%	33.11	65.83	0.14	0.43	0.17	0.25	0.04	0.00	0.04	0.00	100.01	

Electron probe microanalysis instrument; EPMA JCXA-733 and EDS Link860-2; accelerated power supply of 15 kV, probe current of 2×10^{-8}, and excitation beam spot diameter of 0.5 μm; accuracy less than 1%.

[a]*Raw data were extracted with impurity removal to calculate the percentage.*

According to Xie Y, Li S. Discovery of minerals such as pyrite and iron dolomite in fluid inclusions of Taibai Au deposit and its genetic significance. Miner Deposits 2000;19(1):54–60.

4.4.1.2 Sample preparation

Sample preparation is the most important process for achieving high-precision analytical results. The purity of the individual mineral particles should be higher than 98%, to yield better results, and there should be no impurities from the host crystal or others. Minerals in the mature period of formation are usually selected in practical research. The size of the inclusion determines the granularity of the broken samples. Undersized particles lead to the breakage and waste of some inclusions after cracking. It is also important to note that the experiments can be blocked when the particle sizes are too large, thus producing unusable inclusions. In actual operation, the single-mineral sample is generally crushed to particle sizes of 0.2–0.4 mm (equivalent to 40–60 mesh). Thus, the waste of resources can be prevented and the easy opening of inclusions in the mineral can be ensured. To remove contaminants from the surface or cracks in the mineral, the sample is soaked in reagent. For example, quartz is soaked in dilute hydrochloric acid, while calcite is soaked in aqua regia for 1–2 days. Thereafter, the sample is washed with secondary deionized water until the pH of the solution is 7. The sample is filtered, dried, and weighed. In general, 5 g of the sample will burst at 550°C, and then is washed with deionized water for about 4–5 min. The test solution is obtained after 10 min of centrifugal filtration.

4.4.1.3 Sample test

ICP-AES JY-385 and high-resolution ICP-MS ELEMENT were used for the sample test. The concentration of the common heavy metal elements such as Cu, Fe, Pb, Zn, Au, and Ag was 1.0×10^{-9} ppm.

4.4.1.4 Dilution multiples and concentration conversion

The concentration of the heavy metal elements measured in the filtrate does not represent the original ore-forming fluid. This value should be converted, because the measured concentration is the concentration of heavy metal element when the fluid inclusion solution has been diluted 1000 times. The water quality or capacity in the sample should be used for the concentration conversion. Since the density of water $\rho \approx 1$ g/cm^3, 1 mg of water has the equivalent mass of 1 μL of water. Therefore, the dilution factor x is expressed as follows:

$$X = V \cdot \rho \cdot 1000/(M_1 \cdot m_2/m_1) = V \cdot \rho \cdot m_1/(M_1 \cdot m_2) \times 1000,$$

where m_1 is the mass of the sample during the gaseous phase analysis of the inclusions, m_2 is the mass of water in m_1, M_1 is the mass of the sample used for extraction of the filtrate measured, and V is the total volume of the filtrate measured.

It is denoted that C_1 is the concentration of the heavy metal elements in the measured filtrate ($\times 10^{-6}$); C_0 is the concentration of the heavy metal elements in the original ore-forming hydrothermal fluids ($\times 10^{-6}$). With a known dilution factor, C_0 can also be calculated by the following formula:

$$C_0 = X \cdot C_1 = V \cdot \rho \cdot m_1 \cdot C_1 / (M_1 \cdot m_2) \times 10^3.$$

Using this method, the test results obtained for the heavy metal elements in the Tianmashan S—Au deposit and the Datuanshan Cu deposit in primary ore-forming fluids are shown in Table 4.8.

It is found that the fluid inclusions in the deposits are rich in heavy metal components such as Zn, Pb, Au, and Cu. The content of the metal components fluctuates greatly in the range of 10^{-3}—10^{-6}. In the Tianmashan S—Au deposit, the ore-forming fluids are rich in Zn and Fe in the oxide stage due to the high temperature and pressure conditions. In the sulfide stage, the physical and chemical conditions changed with the decrease in temperature. The Cu and Zn content decreased, with increasing Fe content. In the carbonate stage, the low content of Fe indicates a small amount of pyrite. Meanwhile, the increase in Cu and Zn content indicates that Cu and Zn are activated. In the carbonate and sulfide stages, the Ag content is both the highest and the lowest, respectively. The Ag content reflects the Au with similar properties. According to this analysis, the sulfide stage is the main ore-forming period of the Tianmashan S—Au deposit.

Among the various heavy metal elements (except for Ag) at the quartz sulfide stage in ore-forming fluids, the Tianmashan S—Au deposit is generally 1—2 orders of magnitude lower than the Datuanshan Cu deposit. The main ore-forming element Cu has a content of 400×10^{-6} in the ore-forming fluids, while the content of Fe is above 1000×10^{-6}. This accelerates the precipitation of Cu and the formation of Cu deposits, laying the foundation for lots of Au and Ag in the Datuanshan Cu deposit. In fact, Au and Ag are the main components of the Datuanshan Cu deposit. In the Cu deposit, the Au grade is about 0.4—0.9 g/t, with an Ag grade of about 10—15 g/t. In the Cu concentrate, the Au grade is 8 g/t, and the Ag grade is 182 g/t.

4.4.2 Metal components in fluid inclusions in Cu—Au mineral

Au and Cu are exploited in the Dongguashan Cu—Au deposit, Tongling, Anhui, China. The Cu content in this deposit is about 9400 t, with a grade of 10%. The Au content is 22 t, with a grade of 0.24 g/t. The growth layer controls the skarn and the porphyry ores. The primary fluid inclusions of the quartz are conducted with inductively coupled plasma mass spectrometry (ICP-MS) analysis at the stages of potash feldspathization, skarnization, and early and late quartz sulfide. The analytical methods are described as follows.

Based on the observation of the mineral sheets and petrography, 12 samples in four ore-forming stages were selected from the Dongguashan deposit for crushing and washing, thus obtaining clean quartz. This process was immediately accompanied by the analysis of the trace elements in the

TABLE 4.8 Contents of heavy metal elements in primary ore-forming fluids (10^{-6}).

Deposit	Ore-forming stage	Sample nnumber	*W* (Zn)	*W* (Pb)	*W* (Au)	*W* (Cu)	*W* (Fe)	*W* (Ag)
Tianmashan S–Au deposit	Oxide	TM-25/3-4-2	0.68	<7.5	<5.36	41.27	23.58	8.04
	Sulfide	TM-5/4-5-2	0.80	<5.61	<4.01	20.45	32.88	8.42
	Sulfide	TM-55/3-2	<1.37	<9.56	<6.83	28.69	45.08	13.66
	Carbonate	TM-95/38-6	3.31	<5.47	<3.91	62.56	16.81	6.26
Datuanshan Cu ddeposit	Quartz sulfide	DT-460/23-6	59.51	4.91	<3.27	114.45	1023.18	10.79
	Quartz sulfide	DT-460/25-3	135.30	<5.31	<3.79	268.17	1684.66	12.89
	Quartz sulfide	DT-460/25-4	201.70	<5.61	<4.01	433.88	1960.49	14.84

W, mass concentration.

inclusions. In this procedure the particle size of the quartz deposit required was 40–60 mesh, with a purity of 99%.

First, the purified quartz was immersed in dilute hydrochloric acid at 80°C for 1 h. After standing overnight, the acid was removed for drying. Then, 3 g of the sample was weighed. To avoid interference from other inclusions, the sample was broken at 100°C. Then, the temperature was raised to 400°C. After 15 min, the sample was cooled, and 1×10^{-9} of Rh and 3 mL of 5% HNO_3 solution were added for 15 min of ultrasonic oscillations. After centrifugation, the sample was stored for use. The sample was then placed on an ELEMENT-type ICP-MS apparatus manufactured by Finnigan MAT for analysis. Table 4.9 shows the composition of the metal elements in the fluid inclusions of quartz at different stages of the Dongguashan deposit.

It was found that the metal elements Cu, Pb, Zn, Mo, Bi, and Sb are high in the metallogenic stage of the Dongguashan deposit, so these metal elements are well enriched in the inclusions. The highest Cu content is up to 1776.48×10^{-9}, with an average of 0.32×10^{-6}. The highest Pb content is up to 177.64×10^{-9}, with an average of 66.85×10^{-9}. The Zn content is 109.47×10^{-9}, with an average of 57.69×10^{-9}. By analyzing the trace elements of the fluid inclusions, we can obtain information regarding the ore-forming fluid, which is a mineral-enriched hydrothermal fluid. The change in the ore-forming element content in the hydrothermal fluid is consistent with the mineralization degree and grade of ore change in the same stage. The early sulfuration stage is the best stage of mineral concentration. In the late quartz sulfide stage, there may be a mixture of shallow hydrothermal fluids.

4.4.3 Metal components in fluid inclusions of Pb–Zn mineral

The Lanping/Jinding Pb–Zn deposit in Nujiang, Yunnan Province, is a famous enormous deposit of Pb–Zn in China and also the world's oldest sedimentary-type Pb–Zn deposit. The ore consists of sandstone and breccia, with some gangue of the Jinding mining area, where more than 30 primary minerals have been found. There are many common sulfides such as pyrite and sphalerite in the primary mineral, with a small amount of marcasite and chalcopyrite. The particle size of these sulfides ranges from 0.05 to 0.1 mm; the galena is mostly coarse (about 1 mm). Massive coarse crystals exist in the late hydrothermal liquid gangue, such as celestine and pyrite. The quartz is mostly fragmented in a single state, with a small part formed early during mineralization. The sulfides are generally combined with gangue.

The research on primary fluid inclusions of sphalerite and the gangue coexisting with the minerals in the Jinding mine shows that the inclusions are mostly elliptical. Others are irregular in shape, with an average size of 8 μm (2–16 μm). The inclusions are usually in the gaseous and liquid phases. The gas–liquid ratio is between 10% and 80%, with an average of 43%. Component analysis of trace elements of the individual mineral fluid

TABLE 4.9 Trace element composition of fluid inclusions in quartz at different stages of Dongguashan deposit ($\times 10^{-9}$).

Ore-forming stage:	Potash feldspathization		Skarnization		Early quartz sulfide				Late quartz sulfide				
Sample nnumber	1	2	3	4	5	6	7	8	9	10	11	12	Chinese continental crust
Li	2.56	11.74	4.58	42.25	39.47	8.31	3.94	2.70	17.47	3.33	14.50	1.86	44
Be	0.21	0.65	0.24	0.24	0.75	2.17	0.18	0.13	1.39	0.23	0.20	0.10	4.4
Se	0.30	0.41	0.12	0.68	1.21	0.35	0.08	0.12	1.14	0.62	0.11	0.14	11
Ti	33.48	32.17	3.60	21.54	20.43	83.89	8.72	5.16	14.62	6.46	2.87	12.35	6600
V	5.40	61.69	1.64	12.51	64.79	7.15	3.59	9.85	30.64	30.55	10.33	9.32	99
Cr	0.82	1.08	0.98	1.29	29.70	10.18	2.11	2.49	5.20	2.29	0.79	1.04	63
Mn	146.68	246.95	203.56	74.06	118.00	756.98	519.44	544.47	412.25	405.54	231.46	323.25	780
Co	0.92	2.61	0.41	0.50	18.06	0.88	0.76	0.59	2.05	1.95	1.75	0.60	32
Ni	0.50	2.91	1.13	1.13	2.80	0.46	0.54	0.73	1.72	5.60	3.04	1.04	57
Cu	83.16	392.53	29.36	58.04	1776.48	203.18	127.43	204.82	200.02	84.42	323.29	394.53	38
Zn	34.26	44.40	43.02	16.06	31.44	70.26	109.33	109.47	57.53	86.40	44.42	45.71	86
Ca	0.74	3.43	0.35	1.24	22.16	7.51	1.41	0.48	6.81	0.87	0.38	0.35	20
Rb	18.01	35.95	16.24	18.59	35.99	243.39	81.19	81.51	20.85	25.23	24.74	18.32	150

Sr	99.62	36.77	85.67	83.56	1903.26	106.04	82.43	115.13	2009.96	54.18	28.56	26.28	690
Y	1.32	8.38	0.27	1.53	1.46	5.31	0.40	0.48	10.58	0.81	0.39	0.63	27
Mo	4.70	495.65	0.17	37.61	3.94	9.12	103.70	304.73	48.11	3.33	350.15	0.35	2
Cd	0.25	0.36	0.35	0.12	0.37	1.49	1.13	1.19	0.60	0.58	0.37	0.46	550
Sn	0.03	0.03	0.03	0.03	0.13	0.05	0.05	0.05	0.03	0.06	0.02	0.01	4.1
Sb	0.46	1.59	2.49	2.56	0.27	1.84	0.79	1.28	3.84	0.69	1.38	0.22	0.15
Ca	8.03	4.26	6.95	1.44	1.54	14.99	23.54	22.99	4.87	7.45	9.37	5.90	3.7
Ba	4.77	0.64	1.14	1.48	55.10	1.30	1.16	3.91	3.19	1.35	1.16	0.18	610
Re	0.00	0.24	0.00	0.01	0.00	0.00	0.02	0.15	0.05	0.00	0.11	0.00	0.0005
Ti	0.35	0.31	0.23	0.12	0.48	2.16	0.91	1.03	0.41	0.26	0.41	0.24	0.61
Pb	25.30	34.10	39.56	13.79	16.32	177.64	90.31	147.46	81.06	101.09	40.85	34.76	15
Bi	5.48	23.97	7.04	1.15	7.54	12.30	12.45	9.77	6.88	7.94	3.55	3.92	0.127
Th	0.47	0.89	3.22	13.70	0.63	0.20	0.06	0.22	54.66	0.78	0.38	0.03	17
U	0.27	1.52	0.23	1.36	2.16	0.16	0.11	0.17	5.67	0.31	0.14	0.08	5.6

According to Xu X, Lu S, Xie Q. Trace element geochemical characteristics and geological significance of fluid inclusions in Dongguashan Cu-Au deposit, Tongling, Anhui Province. Acta Petrol Sin 2008;24(8):1865–1874 inclusions at the different mineralization stages of the Jinding mine was conducted by thermal explosion extraction and ICP-MS (see Table 4.10).

Table 4.10 shows that the quartz inclusions have high contents of Bi, Cu, Pb, and Zn. The galena inclusions are rich in Ba, Sr, Mn, Co, Ni, Cu, Zn, Cd, Sb, and Mo. The pyrite inclusions are rich in Sr and the ore-forming elements (Ti, V, Cr, Mn, Co, Ni, Cu, Zn, Cd, Ti, and Mo). From the results, Mn, Zn, Ti, and Mo have the highest contents. Here, the three mineral inclusions have the characteristics of enriched elements in the ore-forming fluids. It is found that the content of Zn is high in galena fluid inclusions, with the highest value of 2.31×10^{-5}. The content of Zn is 2.91×10^{-5} in the pyrite fluid inclusions. This indicates that the ore-forming fluid of partial sphalerite is inevitably captured during the ore formation of galena and pyrite.

Wilkinson et al. analyzed the contents of the metal elements in fluid inclusions of sphalerite and quartz from the Pb–Zn mines of northern Arkansas and Ireland. The results showed that the Pb content in the sphalerite fluid inclusions was 2 orders of magnitude higher than that in the quartz fluid inclusions. Based on the correlation between the Pb and the Zn contents, it is speculated that the fluid inclusions in the sphalerite have a higher content of Zn (see Fig. 4.2).

4.4.4 Metal components of group inclusions in quartz and pyrite

The Jiaodong Jiaojia type of Au deposit contains pyrite, galena, and magnetite in addition to natural Au and Ag. Some also contain small amounts of unusual minerals such as arsenopyrite and marcasite. The gangues are mainly quartz, dolomite, and barite. ICP-MS was used to analyze the compositions of the trace elements in the pyrite, quartz, and group inclusions from the Jiaojia, Matang, Dongji, and Hongbu Au deposits of Jiaodong. The specific method is described as follows. The selected quartz and pyrite inclusions were washed to remove secondary inclusions at 150°C, after which the samples were ruptured at 500°C for 15 min. After cooling, the samples were mixed with 3 mL of 5% nitric acid solution. The mixture was finally centrifuged by ultrasonic shaking. Rh 11×10^{-9} was used as the internal standard of the leaching solution for computer determination (see Table 4.11).

The results showed that the pyrite and quartz inclusions were enriched with ore-forming elements such as Cu, Pb, and Zn, reflecting the characteristics of the ore-forming fluids. In the quartz inclusions, the highest Cu content was about 8.57×10^{-6}, with an average of 2.69×10^{-6}. The highest Pb content was 1.27×10^{-7}, with an average of 5.84×10^{-8}. The highest Zn content was 9.91×10^{-8}, with an average of 4.81×10^{-8}. The pyrite had a high quantity of heavy metal components. The highest Cu content was 8.99×10^{-6}, with an average of 3.82×10^{-6}. The highest Pb content was 4.23×10^{-6}, with an average of 3.19×10^{-6}. The highest Zn content was 2.85×10^{-7}, with an average of 7.75×10^{-8}.

TABLE 4.10 Trace element composition of single-mineral fluid inclusions in different ore-forming stages of the Jinding mine (except Rare Earth Element [REE]) ($w(B) \times 10^{-9}$).

Mineralization stage	First mineralization stage			Second mineralization stage			Third mineralization stage		
Sample nnumber	**JY15**	**JY16**	**JY17**	**NC14**	**BC54**	**J-23**	**BC60**	**J-7-21**	**NC5**
Host crystal	Quartz	Quartz	Quartz	Galena	Galena	Galena	Galena	Pyrite	Pyrite
Li	11.911	3.968	6.109	4.593	3.646	2.334	2.568	24.356	9.742
Be	1.008	1.007	1.03	0.986	0.997	0.991	1.006	1.005	1.051
Rb	0.865	1.009	3.745	0.113	0.253	-	0.077	0.367	1.163
Sr	6.393	4.592	24,744	11,131.927	201,190.923	298.129	679.174	148.091	214.708
Cs	0.151	0.089	0.163	0.004	0.037	0.036	0.026	0.116	0.197
Ba	714.953	568.627	11,145.542	25.724	44.188	5.571	124.64	51.695	8.601
Sc	0.982	0.982	1.016	1.017	1.048	0.988	1.27	1.106	1.248
Ti	11.623	10.718	11.435	9.272	9.891	8.956	11.489	11.998	27.112
V	0.458	1.229	1.629	0.526	0.693	0.014	2212	5.482	14.833
Cr	3.467	3.152	2.147	21.027	18.111	9.917	13.291	13.709	37.898
Mn	22.138	20.682	21.152	655.328	938.114	459.632	261,004.275	111,907.387	121,437.279
Co	8.467	0.07	1.654	45.98	0.471	1.111	21.675	49.973	33.561
Ni	2.806	3.052	2679	11.297	3.248	1.035	27.359	222.443	16.107
Cu	36.409	1.465	15.143	0.254	43.915	-	47.498	26.031	16.202
Zn	19.314	13.594	14.016	31,545.56	61,429.208	161,584.122	231,269.456	21,361.372	291,331.81
Ga	0.257	0.292	0.42	0.22	0.431	0.008	0.732	3.147	5.884

Continued

TABLE 4.10 Trace element composition of single-mineral fluid inclusions in different ore-forming stages of the Jinding mine (except Rare Earth Element [REE]) ($w(B) \times 10^{-9}$).—cont'd

Mineralization stage	First mineralization stage			Second mineralization stage			Third mineralization stage		
Sample nnumber	JY15	JY16	JY17	NC14	BC54	J-23	BC60	J-7-21	NC5
Cd	0.105	0.058	0.111	109.155	188.642	431.675	11,983.042	23.228	11,658.636
Sn	0.281	0.223	0.155	0.233	0.068	0.174	2.258	0.514	0.146
Sb	273.086	0.606	4.293	0.654	6.902	0.104	276.017	18.013	5.227
Ti	0.917	1.013	2.082	22.864	11.199	14.826	14.724	11,439.265	11,715.567
Zr	0.556	0.342	0.769	0.056	0.245	0.034	1.67	0.493	1.503
Nb	0.088	0.124	0.062	0.002	0.018	-	0.007	0.086	0.077
Mo	2.16	0.845	1.219	69.146	19.216	0.747	55.749	321,168.919	51,838.986
Hf	0.058	0.041	0.056	0.002	0.034	0.002	0.045	0.2	0.045
Ta	0.002	0.001	0.001	0.001	0.013	0.001	0.001	0.031	0.008
W	0.493	1.049	0.609	0127	0.78	0.115	0.889	1.025	0.309
Re	0.006	0.005	0.003	0.166	0.081	0.012	0.031	0.984	23.892
Th	0.052	0.033	0.175	0.035	0.105	0.01	0.225	0.035	0.226
U	0.443	0.517	0.946	4.073	5.572	0.08	0.797	0.204	3.981

According to Zeng R, Xue C, Gao Y. Study on trace elements of ore-forming fluids in Jinding Pb-Zn deposit, Yunnan Province. J Mineral Petrol 2006;26(3):38–45.

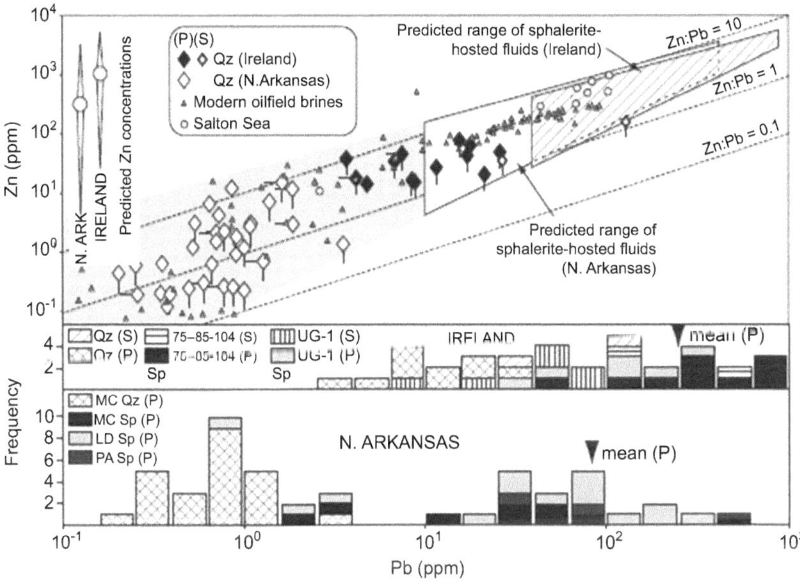

FIGURE 4.2 Contents of metal elements in sphalerite and quartz fluid inclusions of the Pb–Zn mines from northern Arkansas and Ireland. *According to Wilkinson JJ, Stoffell B, Wilkinson CC, et al. Anomalously metal-rich fluids form hydrothermal ore deposits. Science 2009;323(5915):764–767.*

The Xiaoqinling Au mine located on the border of Henan and Shaanxi is a famous Au mineralization zone in China. Most Au quartz veins in this area are mined as a single vein. Typical quartz samples were selected from here for analysis. After that, the trace elements (including rare earth elements) in the inclusions were analyzed for their abundance by ICP-MS. The procedure requires 5 g of quartz sample of +60 to −40 mesh, with a purity greater than 99%. The pretreatment steps of the sample are described as follows. To eliminate the influence of the secondary inclusions, the sample was released by explosion at 100°C once and 400°C for 15 min. After the sample was cooled, 3 mL of 5% nitric acid solution was added and centrifuged for 15 min by ultrasonic oscillation. The clear liquid was poured into a clean instrument for testing. The test instrument was an ELEMENT-type ion mass spectrometer manufactured by Finnigan MAT. Table 4.12 shows the test results.

The results showed that the fluid inclusions in the quartz were rich in various metal components. In addition, the content of metal components was closely related to the symbiotic metal sulfide, namely the host crystal. The concentrations of the metal components differed greatly in the fluid inclusions of the gangue quartz from the different metal sulfides. For the pyrite quartz vein measured by thermal explosion, the contents of Cu, Pb, Zn, and Fe were 1.41×10^{-2}, 3.18×10^{-3}, 4.01×10^{-4}, and 4.10×10^{-2}, respectively. In the

TABLE 4.11 Trace element contents of the quartz and pyrite inclusions ($\times 10^{-12}$).

	Quartz				Pyrite			
	Jch20	Jch59	Mb16	Mb32	Jch20	Jch59	Mb16	Mb32
Li	1046	23,618	19,934	3868	8090	4269	1041	1849
Rb	6045	21,685	26,135	17,749	2557	674	404	432
Ca	149	494	308	297	321	140	50	72
Sr	534,397	182,720	70,914	152,995	11,171	8639	8293	30,549
Ba	1,813,433	236,251	71,650	598,787	2073	137	1010	1773
Y	87	1824	1638	3235	303	4946	3924	1090
Th	57	72	813	498	95	1138	12,728	515
U	30	311	457	227	22	154	2084	95
W	94	24	32	9	64	330	172	65
Sc	16	32	157	32	17	32	46	3
Ti	4493	13,677	18,131	53,994	6719	8015	7651	12,956
V	538	1281	1237	734	249	518	5119	149
Mn	19,355	50,851	84,292	595,241	14,590	40,636	2171	110,818
Co	11,109	1125	20,629	13,474	468,647	8845	92,888	53,964
Ni	8028	7272	4868	1994	173,100	19,996	171,234	6082

Cu	18,883	1,677,969	8,573,670	506,828	727,523	8,998,255	804,168	4,754,765
Zn	32,923	99,175	176,335	42,724	8451	285,378	3309	13,155
Pb	13,566	38,852	127,190	54,089	4,239,588	3,245,663	1,511,135	3,785,021
Ag	58	2674	1200	939	86,487	37,819	3218	29,390
Cd	90	116	54	50	673	913	587	371
Bi	972	4484	8357	3539	3,978,468	301,631	532,787	1,137,482
Ga	513	1196	1796	1105	71	2207	651	173
Au	12	-	-	-	172	379	127	23

According to Li H, Shen Y, Mao J. Study on trace elements in group inclusions in quartz pyrite: a case study of Jiaojia type Au deposit in Jiaodong. Chin J Geol 2004;39(3):320–328.

TABLE 4.12 Trace element content of quartz fluid inclusions in Xiaoqinling Wenyu-Dongchuang Au deposit (except Rare Earth Element [REE]) ($\times 10^{-6}$).

Sample characteristics	DC-10-2 V507 2110m Pyrite quartz vein (I)	DC-17-2 V507 1940m Pyrite galena quartz vein (II)	WY1687-9 V505 1687m Polymetallic sulfide quartz vein (III)	WY1584-4 V507 1584m Coarse galena quartz vein (IV)
Li	55.32	40.95	18.54	34.30
Ti	52.49	36.39	50.87	7.70
V	0.22	0.96	4.87	0.23
Cr	5.06	20.89	7.81	1.84
Mn	143.38	164.54	104.0	33.88
Fe	41,085	16,558	63,295	663.0
Co	25.06	8.18	43.38	0.0658
Cu	14,168	3694	26,971	120.1
Zn	401	88.07	551.2	53.52
Sr	244.1	190.2	306.9	232.2
Y	0.130	0.320	0.369	0.0542
Zr	0.104	0.132	0.60	0.028
Nb	0.0078	0.0058	0.147	0.0068
Mo	17.64	7.87	1.10	0.47
Rh	216.97	295.7	1341.4	91.66
Cs	29.75	35.59	32.27	27.42
Ba	53.31	47.13	89.46	113.67
W	13.34	4.41	5.11	2.05
Pb	3189	560.5	1115.3	94.1
Bi	1.614	1.178	2.66	0.140
Th	0.073	0.056	0.045	0.038
U	0.0301	0.065	0.088	0.011
Mo/W	1.323	1.788	0.215	0.229

TABLE 4.12 Trace element content of quartz fluid inclusions in Xiaoqinling Wenyu-Dongchuang Au deposit (except Rare Earth Element [REE]) ($\times10^{-6}$).—cont'd

	DC-10-2	DC-17-2	WY1687-9	WY1584-4
	V507 2110m	V507 1940m	V505 1687m	V507 1584m
	Pyrite quartz vein	Pyrite galena quartz vein	Polymetallic sulfide quartz vein	Coarse galena quartz vein
Sample characteristics	(I)	(II)	(III)	(IV)
Cu/Zn	35.269	41.944	48.931	2.243
Pb/Zn	7.938	6.364	2.023	1.759
Pb/Cu	0.225	0.152	0.041	0.784
Sr/Ba	4.579	4.037	3.431	2.043

According to Xu J, Xie Y, Liu J. Trace elements and genesis significance of fluid inclusions in Wenyu-Dongyu Au deposit, Xiaoqinling. Geol Prospect 2004;40(4):1–6.

pyrite galena quartz vein, the contents of Cu, Pb, Zn, and Fe were 3.69×10^{-3}, 5.60×10^{-4}, 8.80×10^{-5}, and 1.65×10^{-2}, respectively. In the polymetallic sulfide quartz vein, the contents of Cu, Pb, Zn, and Fe were 2.69×10^{-2}, 1.11×10^{-3}, 5.51×10^{-4}, and 6.32×10^{-2}, respectively. In the coarse galena quartz vein, the contents of Cu, Pb, Zn, and Fe were 1.20×10^{-4}, 9.41×10^{-5}, 5.35×10^{-5}, and 6.63×10^{-4}, respectively.

References

[1] Liu W, Zheng R. Research on gaseous phase composition of inclusions in Huayu Pb-Zn deposit—organic mineralization of MVT deposits (II). Acta Sedimentol Sin 1999;17(4):508–614.

[2] He Y, Wen C, Liu X. Analysis of liquid phase composition of inclusions in multi-non-copper deposits. Metal Mine 2013;(3):108–10.

[3] Xu T, Li Z. Fluid inclusion characteristics and source of ore-forming fluids in Xihuashan tungsten deposit, Jiangxi Province. Resour Surv Environ 2013;34(2):95–101.

[4] Xie Y, Li S. Discovery of minerals such as pyrite and iron dolomite in fluid inclusions of Taibai Au deposit and its genetic significance. Miner Deposits 2000;19(1):54–60.

[5] Xu X, Lu S, Xie Q. Trace element geochemical characteristics and geological significance of fluid inclusions in Dongguashan Cu-Au deposit, Tongling, Anhui Province. Acta Petrol Sin 2008;24(8):1865–74.

[6] Zeng R, Xue C, Gao Y. Study on trace elements of ore-forming fluids in Jinding Pb-Zn deposit, Yunnan Province. J Mineral Petrol 2006;26(3):38–45.

[7] Wilkinson JJ, Stoffell B, Wilkinson CC, et al. Anomalously metal-rich fluids form hydrothermal ore deposits. Science 2009;323(5915):764–7.
[8] Li H, Shen Y, Mao J. Study on trace elements in group inclusions in quartz pyrite: a case study of Jiaojia type Au deposit in Jiaodong. Chin J Geol 2004;39(3):320–8.
[9] Xu J, Xie Y, Liu J. Trace elements and genesis significance of fluid inclusions in Wenyu-Dongyu Au deposit, Xiaoqinling. Geol Prospect 2004;40(4):1–6.

Further reading

[1] Li X, Du Y, Du X. Analysis and calculation of heavy metal elements in ore-forming fluids, vol. 2. Geoscience; 1998.
[2] Gu L, Zaw K, Hu W, et al. Distinctive features of Late Palaeozoic massive sulphide deposits in South China. Ore Geol Rev 2007;31(1):107–38.
[3] Li X, Fan H, Hu F. Composition analysis of single fluid inclusion LA-ICP-MS and its application in ore deposits. Miner Deposits 2010;29(6):1017–28.

Chapter 5

Component release of fluid inclusions in sulfide mineral

Chapter outline

5.1 Analysis of mineral raw materials 99
- 5.1.1 Chalcopyrite and associated minerals 99
- 5.1.2 Sphalerite and quartz 100
- 5.1.3 Galena 102
- 5.1.4 Pyrite 102
  - 5.1.4.1 Pyrite and quartz vein samples from the pyrite deposits in Weixin, Yunnan 103
  - 5.1.4.2 Pyrite and quartz vein samples of polymetallic sulfide deposits from the Dapingzhang area of Yunnan 104

5.2 Research methods for fluid inclusions in sulfide mineral 105

5.3 Morphology and component release of fluid inclusions in chalcopyrite 107
- 5.3.1 Infrared optical microscopic analysis of the fluid inclusions in chalcopyrite 107
- 5.3.2 SEM/EDS of the position of fluid inclusions in chalcopyrite 108
- 5.3.3 HRXMT analysis 111
- 5.3.4 Component release of the fluid inclusions in chalcopyrite 111

5.4 Component release of the fluid inclusions in associated minerals of chalcopyrite 114
- 5.4.1 Component release of the fluid inclusions in the associated quartz and calcite of chalcopyrite 114
- 5.4.2 Component release of fluid inclusions in chalcopyrite, sphalerite, and associated minerals 118

5.5 Morphology and component release of fluid inclusions in sphalerite and associated quartz 123
- 5.5.1 Morphology and types of fluid inclusions in sphalerite and associated quartz 123
- 5.5.2 Freezing point and salinity value of the fluid inclusions in sphalerite 126
- 5.5.3 SEM/EDS analysis of the fluid inclusions on sphalerite surface 127
- 5.5.4 Component release of fluid inclusions in sphalerite and quartz 129

5.6 Morphology and component release of fluid inclusions in galena 133
- 5.6.1 Infrared optics and SEM/EDS of fluid inclusions in galena 133

Fluid Inclusion Effect in Flotation of Sulfide Minerals. https://doi.org/10.1016/B978-0-12-819845-2.00005-3

5.6.2 Component release of inclusions in galena 134

5.7 Morphology and component release of fluid inclusions in pyrite 134

5.7.1 Infrared optical microscopic analysis of fluid inclusions of pyrite in Weixin 134

5.7.2 SEM/EDS detection of the positions of inclusions in pyrite from Weixin 137

5.7.3 High-resolution X-ray topography of microfault pyrite from Weixin 140

5.7.4 Component release of the fluid inclusions of pyrite from Weixin 141

5.7.5 Component release of quartz fluid inclusion in the pyrite deposits of Weixin 143

5.7.6 Component release of fluid inclusions of pyrite from the polymetallic sulfide deposits of Dapingzhang 145

5.7.7 Component release of the fluid inclusions of quartz from polymetallic sulfide deposits of Dapingzhang 147

Further reading 149

It is difficult to obtain a single qualified concentrate in the separation flotation of polymetallic sulfide minerals due to the various concentrations of intermetallic components in the inclusions of polymetallic sulfide. One of the important reasons is that there are a large number of unavoidable metal ions (Cu^{2+}, Pb^{2+}, Zn^{2+}, Ca^{2+}, and Mg^{2+}) that are released into the ore pulp solution. Such metal ions cause nonselective activation of target minerals and gangue, thereby reducing the flotation selectivity of sulfide mineral. In addition, a precipitate is generated under the joint effects of some metal ions in the ore pulp solution and the collector, resulting in an increase in the amount of collector. In terms of such unavoidable ions, it is generally believed that metal ions in a flotation ore pulp solution are derived from oxidative dissolution, the grinding medium, and water floating on the mineral surface.

The previous chapters discussed the formation mechanism of mineral fluid inclusions and composition of the internal material of inclusions. Thus, the metal ions in a slurry solution can be derived from the component release of mineral fluid inclusions. Polymetallic sulfide mineral and its gangue minerals from different regions were taken as research objects in this chapter. We illustrated the opening and component release of the mineral fluid inclusions during the grinding process. Meanwhile, morphology research was performed for the fluid inclusions in the minerals, with concentration analysis of the components.

According to fluid chemistry in mineral formation, the fluid inclusions of metal sulfide minerals contain one of the main components in mineralization. It is a homonymous heavy metal ion. For example, the fluid inclusions in copper sulfide mineral contain Cu, while fluid inclusions in sphalerite contain Zn. It was also shown that there is a close concomitant relationship between minerals and the chemical diversity of ore-forming fluids during mineralization. Thus, components of different host crystals, and those of the host crystals and the gangue minerals, may be captured by one another to form fluid inclusions. In other words, fluid inclusions in copper sulfide minerals may contain components of sphalerite. Fluid inclusions in sphalerite may contain components of copper sulfide. Gangue minerals may contain components of copper sulfide or sphalerite, or a combination of multiple chemical components. Gangue minerals account for the majority of polymetallic sulfide ore, so the influence of gangue mineral inclusions needs to be considered in this work. Therefore, it is necessary to value the effects of the component release of the mineral inclusions on the chemical properties of the slurry solution and surface properties of the mineral.

For this chapter we selected sulfides of Cu, Pb, and Zn, as well as their closely associated gangue minerals, from typical deposits in different areas of Yunnan. The morphology, type, distribution, component release, and concentration of the mineral inclusions were analyzed.

5.1 Analysis of mineral raw materials

5.1.1 Chalcopyrite and associated minerals

Chalcopyrite and its associated minerals were taken from the Dongchuan copper mine in Yunnan, China. The impurities and gangue minerals were manually removed to prepare single mineral of high-purity chalcopyrite. The prepared sample was soaked in 1% sulfuric acid solution for 12 h, and then naturally dried after repeated rinsing with deionized water. The minerals were chemically analyzed to check the purity. According to Table 5.1, the chalcopyrite has high purity, with a small amount of SiO_2 and CaO.

The crystal structure and purity of the chalcopyrite pure mineral were further confirmed by X-ray diffraction (XRD) analysis. Fig. 5.1 shows the diffraction results of the chalcopyrite powder.

Fig. 5.1 shows that the four peaks of XRD are located at 29.34, 48.68, 49.04, and 57.84 degrees, respectively, which is consistent with the data values in the Joint Committee on Powder Diffraction Standards (JCPDS) library (No. 37-0471) and crystal (112), (220), (204), and (312) surfaces. The lattice parameters of chalcopyrite can be obtained from these data, indicating that chalcopyrite has a tetragonal structure, with the space group *I−42d*. The diffraction absorption peaks have high intensity and small width of a half-peak, thus demonstrating that there is a uniform internal structure with an orderly arrangement of particles in chalcopyrite crystals. No other obvious impurities are found in the XRD analysis, which further proves that the prepared chalcopyrite single mineral has a high purity.

TABLE 5.1 Chemical composition of the chalcopyrite.

Element:	Cu	Fe	S	SiO_2	CaO
Content (%)	33.24	27.22	34.81	3.23	1.50

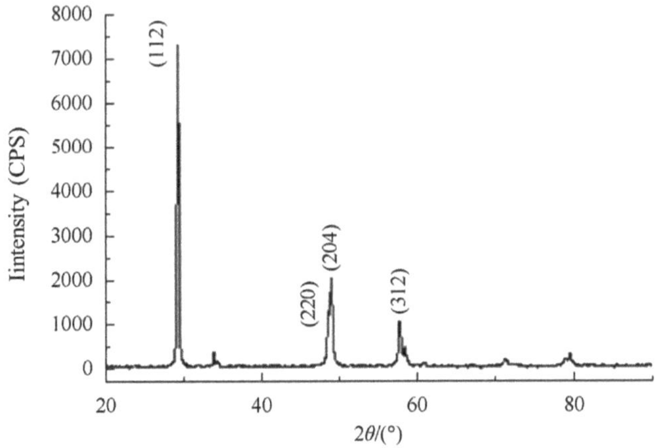

FIGURE 5.1 X-ray diffraction pattern of the chalcopyrite.

TABLE 5.2 Chemical composition of quartz and calcite.

	Chemical composition	Content (%)
Quartz	SiO_2	95.71
Calcite	CaO	53.53

The chalcopyrite crystals are surrounded by bornite, quartz, or calcite crystals. Table 5.2 shows the chemical composition of some associated minerals. Consistent with the theoretical values, the percentages of Cu, Fe, and S in the bornite are 63.01%, 11.01%, and 25.23%, respectively. Converted to atomic ratio, their ratio is close to the ideal stoichiometric number of 5:1:4. The contents of SiO_2 in the quartz and CaO in the calcite are also close to their theoretical values. Thus, the bornite, quartz, and calcite have a high purity.

5.1.2 Sphalerite and quartz

The mineral raw material was collected from the Huize Pb–Zn mine in Yunnan Province. The sphalerite is a pure massive mineral associated with blocky gangue, which mainly consists of quartz of high crystal purity and small amount of calcite crystals. High-purity sphalerite crystals were obtained after artificially eliminating the associated gangue minerals such as quartz and calcite. The prepared sample was soaked in 1% sulfuric acid solution for 12 h, and thereafter repeatedly rinsed with deionized water and dried naturally. After that, the high-purity sphalerite was subjected to multielement analysis and XRD (D/Max 2200, Rigaku, Japan) analysis (see Table 5.3 and Fig. 5.2).

TABLE 5.3 Chemical composition of sphalerite.

Element:	Zn	S	Fe	Pb	SiO_2
Content (%)	64.84	32.81	0.19	0.014	1.21

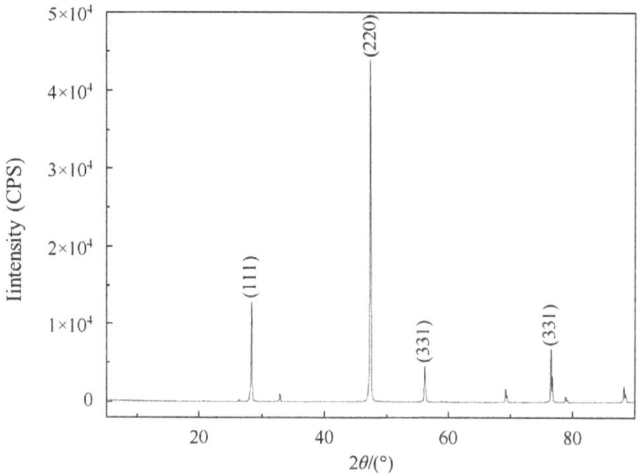

FIGURE 5.2 X-ray diffraction of sphalerite crystal.

The theoretical Zn content of sphalerite is 67.07%, and according to Table 5.3, that of the sphalerite raw material was 64.84%. The purity of the sphalerite obtained was about 96.68%, indicating that the purity of the mineral meets the requirement of the analysis.

Sphalerite has two kinds of isomeride: cubic-structure sphalerite (β-ZnS) and fiber sphalerite (α-ZnS). Fig. 5.2 shows that the four peaks of sphalerite crystal diffraction appear at 28.56, 47.52, 56.29, and 76.81 degrees, respectively, which is consistent with the XRD card of standard sphalerite (No. 050566). In addition, the four diffraction peaks correspond to sphalerite (111), (220), (331), and (331) surfaces, respectively, indicating that the selected sphalerite is consistent with standard sphalerite. For standard sphalerite, the crystal parameters are an equiaxed cubic system, namely, $a = b = c = 5.414$ Å, and $\alpha = \beta = \gamma = 90$ degrees, while those of the selected sphalerite are $a = b = c = 5.406$ Å, and $\alpha = \beta = \gamma = 90$ degrees. The XRD results show that the diffraction absorption peaks have high intensity and small half-peak width, indicating that the sphalerite has coarse crystallites, with uniform internal structure of crystal and an orderly arrangement of atoms. No other impurity minerals were found by XRD analysis.

5.1.3 Galena

The galena was also collected from the Huize Pb–Zn mine in Yunnan Province. High-purity mineral crystals were obtained after successively removing the associated gangue minerals such as calcite and quartz. The prepared sample was immersed in 1% sulfuric acid solution for 12 h. After repeated rinsing with deionized water, the sample was naturally dried for use. Multielement analysis and XRD analysis were performed on the high-purity galena, with the results shown in Table 5.4 and Fig. 5.3.

The multielement analysis and XRD analysis results of single-mineral galena indicate that the mineral obtained had a high purity. There were no other impurities except for small amounts of iron, zinc, copper, and silicon, and the purity achieved was about 98.13%.

5.1.4 Pyrite

Pyrite is one of the most common sulfide minerals and can be found in almost all metal sulfide ore deposits. Therefore, the amount of fluid inclusions in the pyrite accounts for a great proportion of the total number of inclusions in the whole deposit. Pyrite and gangue minerals were selected from the mineral deposits in two different regions of Yunnan Province as follows.

TABLE 5.4 Chemical composition of galena used in experiments.

Element:	Pb	S	Fe	Zn	Cu	SiO_2
Content (%)	84.71	13.42	0.33	0.04	0.015	1.32

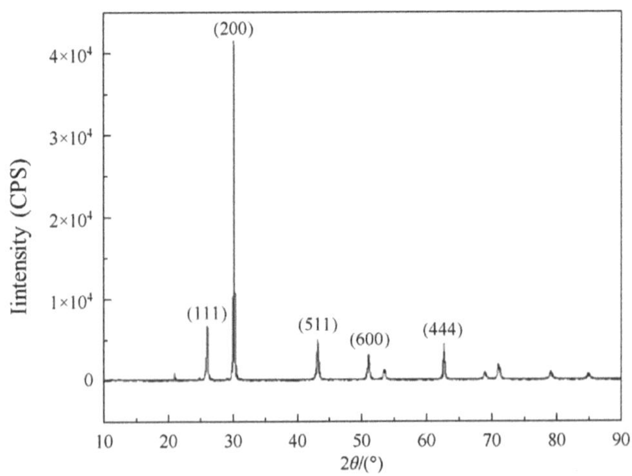

FIGURE 5.3 X-ray diffraction pattern of the galena.

5.1.4.1 *Pyrite and quartz vein samples from the pyrite deposits in Weixin, Yunnan*

This pyrite deposit is a typical hydrothermal metallogenic large deposit, of which the surface is spray shaped and the deep part is immersion shaped. The ore belt is distributed in the northwest direction as an inverted S shape. It starts from the Dawan sulfur plant in Gong County, Sichuan Province, through Shunhe–Gaotian–new Zhenxiong field–Heishuzhuang in Weixin, Yunnan, to Dapo–Shiniujiao–Maochang of Dafang in Guizhou. The belt has a total length of about 175 km and a width of 10–20 km. The main ore belt is distributed in the tip-out section of the eastern Emeishan basalt. The ore body of pyrite occurs in the tuff of the bottom erosion surface of the Maokou limestone (P1m) (some belongs to the flint layer). It presents a podlike and stratiform shape, controlled by undulation of the floor karst erosion surface. The thickness generally ranges from 1 to 5 m, with a maximum value of 7.19 m. It is distributed 10–30 degrees in the northwest direction, inclined to the northeast. The quartz ore belt is interspersed in the pyrite ore body. The ore is mainly composed of pyrite and quartz, as well as a small amount of marcasite, chalcopyrite, and siderite. The geological data show that the sulfur grade of the deposit ranges from 15% to 22%, with an average of 19.37%. Contents of other elements are as follows: Fe, 10%–15%; SiO_2, 50%–65%; TiO_2, 2%–6% (rutile); Cu, 0.05%–0.5%; CaO, 0.1%–1.0%; MgO, 0.1%–0.5%; MnO, 0.01%–0.03%; F, 0.03%–0.05%; As, <0.5%; Pb, <0.01%; and Zn, <0.01%.

Pure lump ores from the pyrite veins and quartz veins were used as raw materials, crushed to 0.5–1 mm, and pure pyrite was picked from the pyrite vein as the pyrite sample and pure quartz was selected from the quartz vein as the quartz sample. The collected samples were immersed in 1% sulfuric acid solution for 12 h. After repeated rinsing with deionized water, they were naturally dried for the chemical analysis. Table 5.5 shows that the pyrite was of high purity. The ratio of sulfur to iron was close to 2:1, with a small amount of SiO_2 and Cu.

XRD analysis was performed on the selected pyrite single-mineral raw materials (see Fig. 5.4). The results showed that the pyrite obtained from this

TABLE 5.5 Chemical composition of the samples from the Weixin area.

	Element content (wt%)					
Sample name	Fe	S	SiO_2	Cu	Pb	Zn
Pyrite	45.84	53.81	0.21	0.014	<0.01	<0.01
Quartz	0.57	<0.05	99.2	<0.01	<0.01	<0.01

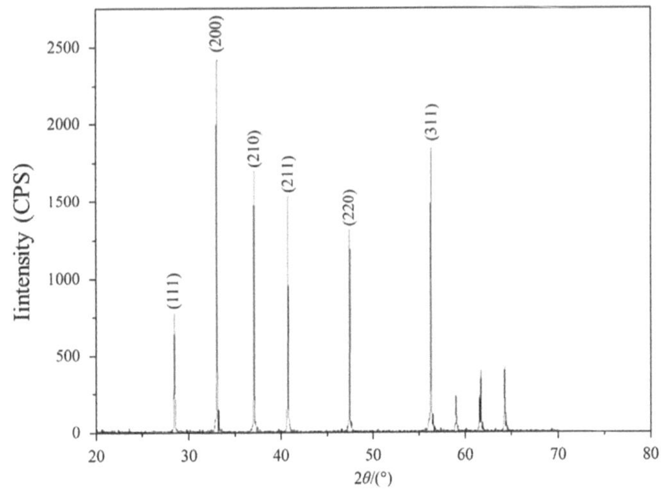

FIGURE 5.4 X-ray diffraction pattern of the pyrite.

process had a purity of over 99%, without any obvious impurities. Each diffraction peak of the sample was matched with that of pyrite in the JCPDS database (IDCC10710053). Corresponding to pyrite (111), (200), (210), (211), (220), and (311) surfaces, six strong peaks of XRD were located at 28.517, 33.034, 37.0753, 40.750, 47.403, and 56.261 degrees, respectively. The lattice constant for pyrite was a = b = c = 5.424 Å, and $\alpha = \beta = \gamma = 90$ degrees. The mineral sample had an equiaxed cubic structure, with a molecular formula of FeS_2 and symmetric atomic structure of T6 h-Pa3. Fig. 5.4 shows the characteristic diffraction peaks of the pyrite with high intensity and small half-peak width, which indicate that the ideal pyrite crystal was obtained, with uniform internal structure and orderly arrangement of internal particles.

5.1.4.2 Pyrite and quartz vein samples of polymetallic sulfide deposits from the Dapingzhang area of Yunnan

The pyrite vein and quartz vein samples from the Dapingzhang area were produced from the copper–lead–zinc–sulfur polymetallic sulfide deposit, which is in the town of Dapingzhang, Lantsang area, Pu'er, Yunnan Province. A few pure lump ores from the pyrite veins and quartz veins were used as raw materials and crushed to 0.5–1 mm. Pure pyrite was picked from a pyrite vein as the pyrite sample, and pure quartz from a quartz vein was selected as the quartz sample. The prepared sample was immersed in 1% sulfuric acid solution for 12 h. After being repeatedly rinsed with deionized water, it was naturally dried for chemical analysis. Table 5.6 shows that the pyrite was of high purity. The ratio of sulfur to iron was close to 2:1, with a small amount of SiO_2 and Cu. Quantitative analysis by XRD indicated that the purity of the pyrite and quartz samples was over 99%, without any obvious impurities.

TABLE 5.6 Chemical composition of the samples from the Dapingzhang area.

	Element content (wt%)					
Sample name	Fe	S	SiO_2	Cu	Pb	Zn
Pyrite	44.78	52.77	<0.05	0.034	<0.01	<0.01
Quartz	0.5	<0.05	99.2	<0.01	<0.01	<0.01

5.2 Research methods for fluid inclusions in sulfide mineral

In this chapter, the following major topics are involved in the study of fluid inclusions in Cu–Pb–Zn sulfide mineral and their associated minerals: the detection and morphology of inclusions in the minerals and the release of the inclusion components and study of their concentrations.

It is known that most metal sulfides are opaque under optical microscopy, especially the common nonferrous metals of Cu–Pb–Zn sulfide. Ordinary optical microscopes are limited to studies of transparent (e.g., quartz) and translucent (e.g., sphalerite) minerals that are symbiotic with metal sulfides. Therefore, the combination of ordinary optical microscopy with the advanced infrared–ultraviolet (IR–UV) optical imaging technology was adopted for the detection of fluid inclusions in minerals and the study of their morphology. To achieve a better view, first of all, the block mineral crystals were cut and ground into sheets with thickness of 0.1–0.3 mm. Then, they were observed under a BX51 IR–UV microscope (manufactured by Fluid, Inc., USA). The image obtained was converted into a data signal by infrared electronic induction, outputting the image through computer processing software.

The IR–UV test of the mineral fluid inclusions was completed at the State Key Laboratory for Mineral Deposits Research, Nanjing University. In addition, scanning electron microscopy/energy-dispersive X-ray spectroscopy (SEM/EDS) and high-resolution X-ray microfault three-dimensional imaging (HRXMT) analysis were used to examine the mineral inclusions to different extents. SEM was adopted from Philips XL30ESEM-TM electron microscope. EDAX's Genesis spectrometer was used for the spectroscopic analysis. We used SEM and EDS to detect the surfaces of the original minerals and the damaged areas after the fluid surface inclusions were broken. The differences in geometrical morphology and chemical composition of the surfaces were compared.

During the test, the ground sheets were cut into 1 × 1 cm, and then fixed on a scanning electron microscope. First, they were scanned with a low-power microscope until any opened inclusions were found. Prior to this, microscopic imaging and energy spectrum analysis were performed. The spatial distribution of the fluid inclusions in individual mineral particles was analyzed using HRXMT. Single mineral particles were placed in a closed syringe for HRXMT scanning to obtain high-pixel-resolution images. The sample was scanned at a voltage of about 150 kV for a certain period of time, to obtain sufficient X-ray intensity points. The HRXMT study was performed at the laboratory of the academician J.D. Miller, School of Metallurgical Engineering, University of Utah, Salt Lake City, Utah, USA.

The research on mineral inclusion components can be divided into two categories: component study of group inclusions and component study of individual inclusions. For the discipline of mineral processing, more attention is paid to the component release of mineral inclusions and their concentrations. In addition, we also focus on the changes in the chemical properties of ore pulp solution and properties of the mineral surface caused by these components. The study on components of group inclusions in minerals generally includes the following steps: single mineral selection, cleaning, opening of inclusions, and extraction and analysis of the inclusion component.

The opening of inclusions can be further divided into the mechanical crushing method, the grinding method, and thermal explosion. While thermal explosion is often used in the field of geology, it is not common in mineral processing due to the difficultly in simulating the grinding process and the high cost of operation. Therefore, in this work, a unique grinding method (superclean grinding) with ultrasonic extraction was used to obtain the component liquid of the mineral inclusions. High-precision inductively coupled plasma mass spectrometry (ICP-MS) and inductively coupled plasma light (ICP-AES) were adopted to determine the cationic metal components in fluid inclusions. Liquid-phase ion chromatography (IC) was used to determine the anion components. ICP-MS was used for the detection of parts per billion content in the solution, whereas ICP-AES was used for detection of the parts per million level. These methods were beneficial for their detection advantages, thus improving the detection accuracy. The ICP-MS model (ELAN DRC II) was manufactured by PerkinElmer, USA, while the ICP-AES apparatus (PS1000) was from Leeman Labs, USA. The anions in the solution were detected using a Waters 820 high-performance liquid chromatograph (for IC). All containers used in the experiment had been soaked in 6 mol/L nitric acid solution for 1 week, washed with deionized and ultrapure water, and then dried naturally. The specific test methods were as follows.

Single mineral with particle size of about 0.5−1.0 cm was washed with an ultrasonic oscillator instrument (SY2200, Shanghai, China) 10 times in deionized water, and then naturally dried in an argon gas box (homemade). During the experiment, 2 g of the dried sample was weighed each time and ground with an MM400 ultrapure ball mill (made in Germany) at a frequency of 900 min^{-1} for different times. The grinding device was a 50-mL stainless steel ball mill pot, containing only one stainless steel ball with a diameter of 15 mm. The ground samples were placed in glass centrifuge tubes, and 40 mL of deionized water was added to each tube for washing by ultrasonic oscillation for 1 min. After washing, the centrifuge tubes were placed in a centrifuge (TGL-16, made in Jintan, Jiangsu, China) for high-speed centrifugal solid–liquid separation. Then, the supernatant was taken out for detection of the metal elements' concentrations in solution by ICP-MS/AES, while liquid chromatograph IC was adopted to detect the anion concentration in the same supernatant.

To prevent the effects of oxygen during the experiment, the aforementioned experimental procedures were carried out inside an argon-protected experimental glove box. The test temperature was maintained at 25°C. The deionized water was supplied by an American-made Milli-Q50, providing high-purity water with resistivity of 18 MΩ. A blank control experiment was carried out to ensure the precision of the whole experiment. That is, the mineral was replaced by deionized water to evaluate the contribution of deionized water and grinding medium to the elements Cu, Pb, Zn, Ca, Mg, Cl^-, and SO_4^{2-} in the solution. The results showed that the contents of these components were on the order of magnitudes of 10^{-9} and 10^{-10} mol/L, which indicated that the contribution of water and grinding medium to the chemical components of the entire solution system was negligible.

5.3 Morphology and component release of fluid inclusions in chalcopyrite

5.3.1 Infrared optical microscopic analysis of the fluid inclusions in chalcopyrite

Chalcopyrite is an opaque mineral under visible light. Hence, it is difficult to study the inclusions of chalcopyrite with ordinary optical microscopes. The block-shaped chalcopyrite crystals were cut into sheets with thickness of 0.1–0.3 mm. Only the morphology of the chalcopyrite surface and shallow inclusions could be obtained with a BX51 infrared microscope (see Fig. 5.5). Accurate existence of the fluid inclusions was difficult to determine in some areas; thus other techniques were used to assist the analysis.

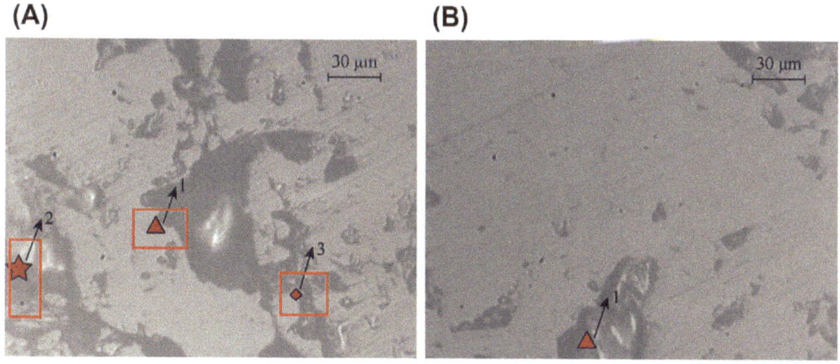

FIGURE 5.5 Infrared optical micrographs of (A) surface and (B) shallow fluid inclusions.

Under an infrared microscope, the suspected fluid inclusions in chalcopyrite sheets were isolated (see Fig. 5.5B) and clustered (see Fig. 5.5A). The shapes were striplike (see Fig. 5.5A, 1), elliptical (see Fig. 5.5A, 3, and B, 1), and irregular (see Fig. 5.5A, 3). They were distributed along the orientation of the crystal growth belt of chalcopyrite (direction of 45 degrees). The sizes of the inclusions ranged from 3 to 60 μm, with lengths of 3–50 μm and widths of 2–30 μm. The colors of the inclusions appeared different from the surrounding area (see Fig. 5.5A, 1 and 2, and B, 1), and they could exist in liquid phase or gaseous phase or a combination of the two. Fig. 5.5A and B show that chalcopyrite has bright and dark streaks at a 45-degree angle, which result from the difference in absorption of infrared light by different atomic states and crystal textures of various elements. The infrared optical microscopy imaging shows that there are a large number of fluid inclusions in this mineral.

5.3.2 SEM/EDS of the position of fluid inclusions in chalcopyrite

Inclusions in chalcopyrite surface and superficial area could be destroyed and opened during grinding of the mineral sheets. SEM observation shows that there were many dimples on the dissociation surfaces of chalcopyrite. Such dimples may be the traces left by the opening of inclusions, called the inclusion domain (see Fig. 5.6A–F). From the structure and morphology of the inclusion domain, it is seen that inclusions vary in size and shape. The sizes range from 5 to 60 μm, or even larger (see Fig. 5.6B). Some of the inclusions are isolated (see Fig. 5.6A, and C–F), while some are concentrated

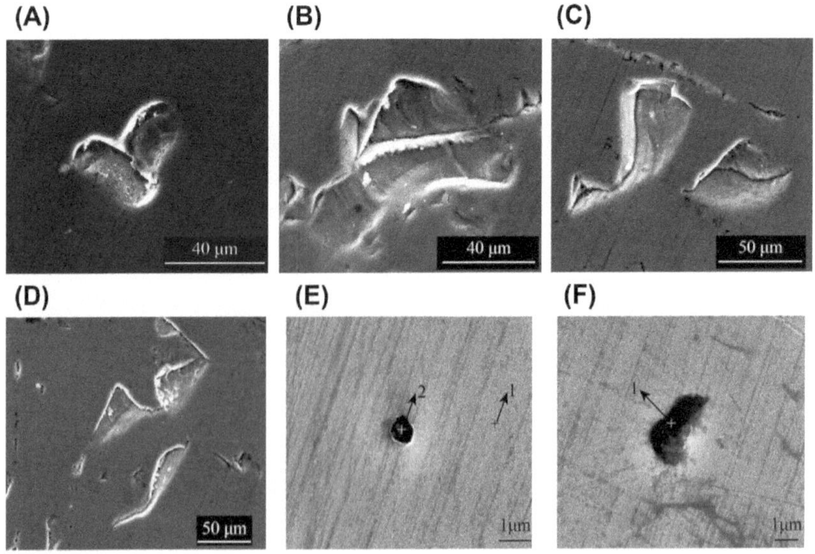

FIGURE 5.6 Scanning electron microscopy morphology of residual position domains. (A)–(F) are selected SEM images.

(Fig. 5.6B). They tend to be strip shaped (see Fig. 5.6D), elliptical (see Fig. 5.6A and F), spherical (see Fig. 5.6E), and irregular (see Fig. 5.6B and C). There are regular stripes at a 45-degree angle, which are crystal growth belts from the hydrothermal mineralization. The inclusions are oriented along the crystal growth region. In terms of structure and state of the inclusions, the result of the SEM topography analysis was consistent with that of the infrared optical microscopy imaging of the surface and shallow fluid inclusions.

Furthermore, EDS was performed on the recessed area, and the results proved again that the dimples on the dissociation surface were residues of inclusions rather than scratches caused by cutting the surface. For comparison, the following areas were taken for analysis: plane area around the dimples (see Fig. 5.6E, 1) and the recessed positions (see Fig. 5.6E, 2, and 5.6F, 1), namely the areas marked with "+" (see Fig. 5.6). Fig. 5.7 shows the EDS spectrum, while Table 5.7 shows the semiquantitative element contents. The electron energy spectrum of Fig. 5.6E, 1, showed only the peaks of S, Fe, and Cu. The results of the semiquantitative analysis showed that the atomic concentration ratio of S, Fe, and Cu is close to 2:1:1, which is the stoichiometric theoretical composition of chalcopyrite. Apparently, the chalcopyrite around the dimples has a high purity and lacks impurities. In the energy spectrum of Fig. 5.6E, 2, peaks of O and Cr appeared in addition to the S, Fe, and Cu. In addition, the semiquantitative analysis results also showed that the atomic concentrations of

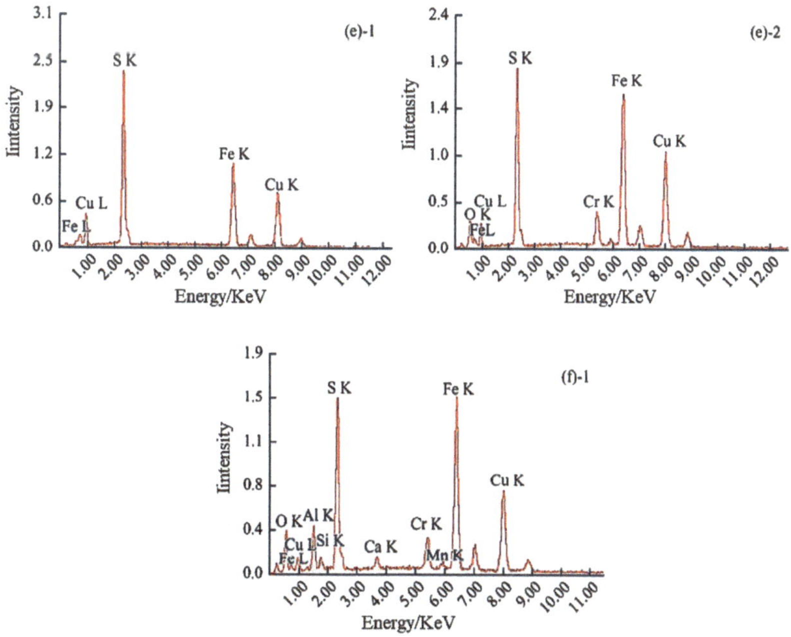

FIGURE 5.7 Energy-dispersive X-ray spectroscopy electron spectrum of the inclusion domain.

TABLE 5.7 **Semiquantitative analysis results of concentration of elements in EDS.**

		Concentration (%)	
	Element	Mass concentration	Atomic concentration
Fig. 5.6E, 1	SK	36.86	52.02
	FeK	30.71	24.88
	CuK	32.43	23.10
Fig. 5.6E,2	OK	09.38	24.14
	SK	21.40	27.48
	CrK	05.18	04.10
	FeK	31.30	23.07
	CuK	32.73	21.21
Fig. 5.6F, 1	OK	07.34	17.22
	SK	17.83	20.87
	AlK	04.54	6.32
	SiK	12.37	16.53
	CaK	06.21	5.81
	MnK	04.19	2.86
	CrK	05.18	3.74
	FeK	20.08	13.49
	CuK	22.26	13.15

O and Cr were close to 30%, with mass percentage of about 15%. The atomic concentration ratios of S, Fe, and Cu far exceeded the theoretical stoichiometric value for chalcopyrite, with the concentrations of Cu and Fe atoms nearly doubling the stoichiometric values, thus indicating that the components of dimples are different from those of the surrounding area, with a large number of other elements and homonymous cations.

In the energy spectrum of Fig. 5.6F, 1, peaks of more elements appear in addition to S, Fe, and Cu, such as peaks of O, Al, Si, Ca, Mn, and Cr. The semiquantitative analysis results showed that the content of such impurity elements was close to 40%, and the atomic concentration ratio of the Cu, Fe, and S was higher than the theoretical value. According to the results of the EDS spectrum and semiquantitative analysis, the dimples were naturally existing rather than the results of cutting the surface. The elements of the phase interface components such as Cu, Fe, Al, Ca, and Cr detected by EDS were inherently generated, possibly resulting from the release of the fluid inclusions. In other words, the dimples were the remaining areas after the inclusions were opened. Various ions were adsorbed on the walls of the inclusions after volatilization of the vapor and liquid inclusions.

5.3.3 HRXMT analysis

HRXMT provides a multidimensional micrometer resolution image of the internal structure of the sample. Thousands of particles can be scanned by HRXMT at the same time, which helps to search for fluid inclusion structures with complete internal holes. HRXMT was used for three-dimensional CT scanning of chalcopyrite particles (see Fig. 5.8A). The HRXMT image revealed gaps between the particles and cracks inside the particles. Especially, consistent with the SEM results, some fluid inclusions inside the chalcopyrite were also detected, proving the existence of fluid inclusions. Fig. 5.8B shows the size of holes in the fluid inclusions (circled in Fig. 5.8A), with a diameter of approximately 37 μm, which is consistent with the results of previous sections. Fig. 5.9 shows the CT scanning image.

5.3.4 Component release of the fluid inclusions in chalcopyrite

In geochemistry, studies on fluid inclusions indicate that the contents of fluid inclusions vary among deposits. Some deposits have more abundant fluid inclusion contents, including gaseous, liquid, and solid phases. The abundance of liquid fluid inclusions accounts for 1%–3%. This work studied Cu^{n+}, Fe^{n+}, Cl^-, and SO_4^{2-} plasmas in solution released from chalcopyrite's fluid

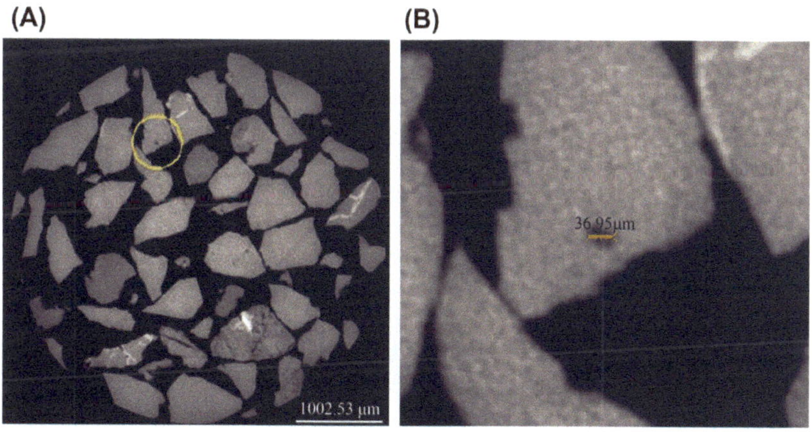

FIGURE 5.8 High-resolution X-ray microfault 3D imaging of chalcopyrite. (A) Overall. (B) Position domain.

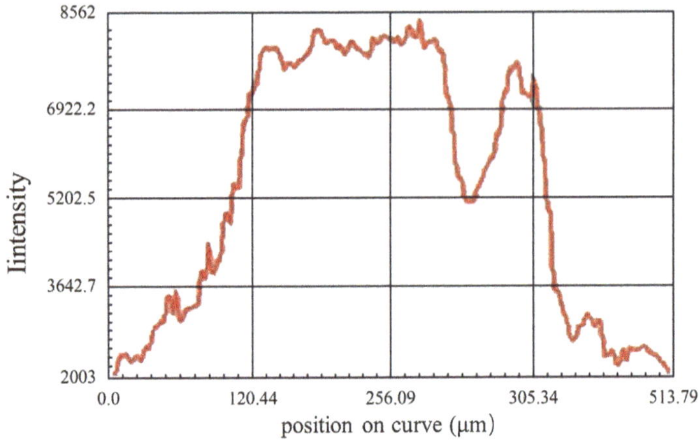

FIGURE 5.9 The CT number distribution of the line in Fig. 5.8B.

TABLE 5.8 Concentrations of Cu, Fe, Cl^-, and SO_4^{2-} released from the chalcopyrite fluid inclusions.

	Concentration ($\times 10^{-6}$ mol/L)			
Grinding time (min)	Cu	Fe	Cl^-	SO_4^{2-}
6	0.16	0.52	34.7	60.68
8	0.21	0.80	41.18	77.86
10	0.45	1.67	77.29	79.21
12	2.14	5.59	94.78	81.71
14	5.79	17.20	107.76	94.51

inclusions, comparing the contributions of fluid inclusions and mineral surface dissolution to unavoidable ions. The cation components in chalcopyrite fluid inclusions mainly included Na^+, K^+, Mg^{2+}, and Ca^{2+}, as well as intrinsic homonymous ions of Cu^{2+} and Fe^{2+}. The anions included Cl^-, SO_4^{2-}, and F^-.

Many researchers in geochemistry focus on the composition and content of these ions. Direct measurement methods have been formed, with good results. However, indirect determination is mainly recommended by geochemical research for the study of heavy metal ions. In this test, the indirect determination method was used to determine the components in chalcopyrite fluid inclusions. Concentrations of Cu and Fe (ICP-MS) in the solution released from mineral inclusions of cracked chalcopyrite were detected by ICP-MS. Cl^- and SO_4^{2-} (IC) were detected by IC of high-performance liquid chromatography (see Table 5.8).

Table 5.8 shows that after 1 min of ultrasonic vibration, the total concentrations of Cu and Fe (C_{CuT} and C_{FeT}) in the solution significantly increased with an increase in the grinding time. Within 6–14 min, C_{CuT} increased from 0.16×10^{-6} to 5.79×10^{-6} mol/L, while C_{FeT} increased from 0.52×10^{-6} to 17.20×10^{-6} mol/L. The values were significantly increased. Most sulfide minerals, including chalcopyrite, are generated by hydrothermal mineralization. The mineralization process is accompanied by contact and interaction with brine. Lots of studies have shown that NaCl, H_2S, CO_2, and sulfate are abundant in fluid inclusions. The existence of these substances can be direct evidence for component release of inclusions. Therefore, the work detected the concentrations of Cl^- and SO_4^{2-} anions released from the chalcopyrite inclusions when they are ruptured, and the results show that the concentrations of Cl^- and SO_4^{2-} in the solution gradually increased as the crushing time increased.

Within the crushing time of 6–14 min, the concentrations of Cl^- and SO_4^{2-} in the solution had increased from 34.7×10^{-6} and 60.68×10^{-6} mol/L to 107.76×10^{-6} and 94.51×10^{-6} mol/L, respectively. Such variation is large in order of magnitude. The magnitude of the concentrations of the cations and anions in Table 5.8 is consistent with the results of previous literature on fluid inclusions. The concentrations of Cl^- and SO_4^{2-} by ICP-MS in blank water are 7.6×10^{-10} and 20.4×10^{-10} mol/L, with 3 orders of magnitude difference from those in the solution, indicating that the test and analysis systems have little effect on the results.

It is necessary to further exclude the possibility that the large amounts of Cu and Fe elements in the solution are derived from dissolution of minerals. Thus, after 14 min of crushing, the mineral particles were washed several times until the concentrations of Cu and Fe in the solution became 9.6×10^{-9} and 25.9×10^{-9} mol/L, respectively. Then, a dissolution test was carried out in a glove box for comparison and analysis of the results. The products after the maximum crushing time of 14 min were selected and placed in a centrifuge tube. Some deionized water was added to the samples, which were cleaned by ultrasonic vibration for 1 min. After that, the sample was centrifuged for solid–liquid separation, adopting the cleaned solid powder as the test material. The purpose was to remove components of Cu and Fe released from the fluid inclusions in the chalcopyrite. The procedure for the dissolution analysis is as follows. Put 2 g of the solid particles and 40 mL of deionized water in a glass reactor of 50 mL volume and magnetically stir for 7 h at a speed of 1000 rpm. Afterward the sample is centrifuged for solid–liquid separation. The supernatant is collected and sealed in a glass vial at room temperature of 25°C for ICP-MS detection. After 7 h of dissolution, we obtained the concentration test results for Cu and Fe in the solution (see Table 5.9).

Table 5.9 shows that the concentrations of Cu and Fe in the solution are 0.05×10^{-6} and 0.12×10^{-6} mol/L, respectively. The values have nearly 2 orders of magnitude difference from the results of Table 5.8. Thus, such large amounts of Cu, Fe, Cl^-, and SO_4^{2-} must have resulted from the release of fluid inclusions in the mineral rather than the dissolution of chalcopyrite.

TABLE 5.9 Experimental total concentration of Cu and Fe by inductively coupled plasma mass spectrometry.

Grinding time/min	Dissolution time/h	pH	C_{CuT} (mol/)L	C_{FeT} (mol/L)
14	7	6.9	0.05×10^{-6}	0.12×10^{-6}

5.4 Component release of the fluid inclusions in associated minerals of chalcopyrite

Mineralization of minerals is accompanied by the interactions of various elements and evolution of different types of crystals, leading to the generation of rich and peculiar multimineral symbiotic combinations. The symbiotic mineral fluid inclusions are studied by geologists for determination of age, nature, and location of ore deposits. The mechanism involves fluid mixing and coupling precipitation. Due to contact, crossover, and dip-dyeing processes of the combined minerals in different periods, interpenetration of fluids occurs between the different mineral crystals. The study of inclusions in associated minerals of chalcopyrite can further confirm the existence of inclusions and component release. In addition, it is also necessary to confirm the existence and component release of inclusions in other metal minerals and gangue minerals that are closely associated with the main mineral. They have important influence on the chemical composition of flotation solution and the flotation process itself.

5.4.1 Component release of the fluid inclusions in the associated quartz and calcite of chalcopyrite

A petrographic study was performed for fluid inclusions in the associated quartz and calcite of chalcopyrite. It was found that the two transparent minerals contain large amounts of fluid inclusions (see Figs. 5.10 and 5.11). The fluid inclusions in calcite are mainly parallel or perpendicular to cleavage or along the microcracks, with a small amount distributed in isolation. Under the microscope clear interpenetration of fluid inclusions at different stages is observed, which could reflect the traces of the three stages. In general, fluid inclusions observed under calcite are small, ranging from a few to a few dozen micrometers, with shapes of negative crystals, strips, ellipses, and irregular forms. At room temperature, the fluid inclusions exist in liquid and gas phases, while the gas/liquid ratio is not completely consistent. In addition to gas and liquid inclusions, the microscope shows a small amount of single-phase inclusions of pure gas phase and pure liquid phase.

Fig. 5.11 shows the association micrograph of quartz and opaque minerals. Quartz is a kind of transparent mineral, so its mineralogical features such as

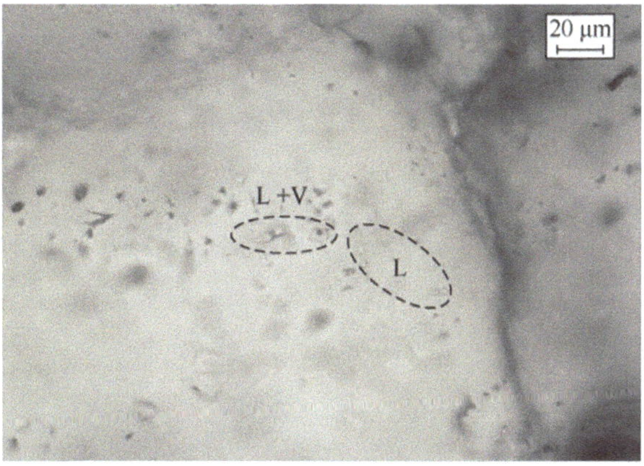

FIGURE 5.10 Optical microscopy imaging of fluid inclusions in calcite. *L*, liquid phase; *V*, vapor phase.

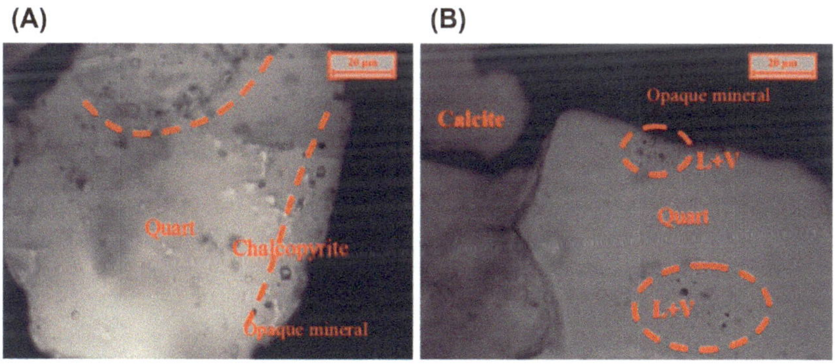

FIGURE 5.11 Association of quartz and opaque minerals micrograph. (A) Slit-shaped and isolated distribution. (B) Isolated distribution. *L*, liquid phase; *V*, vapor phase.

internal structure can be observed under the optical microscope. Clear fluid inclusions are observed in the quartz, in slit-shaped distribution (see Fig. 5.11A) and isolated distribution (see Fig. 5.11B). At room temperature, we can observe the well defined gaseous—liquid phase and the liquid-rich fluid inclusions (see Fig. 5.11A and B). The sizes range from a few to a few dozen micrometers, with shapes of circles and irregular forms.

The integrated methods used for identifying stable isotopes of H, O, S, and C, as well as inclusions in gangue minerals, were adopted to study the properties and sources of ore-forming fluids. It was found that the gangue minerals at the sampling site contained inclusions of daughter minerals (NaCl and KCl) and organic components in addition to the liquid and CO_2. The determination results showed that the homogenization temperature of the diagenetic deposits was about 150°C, with the salinity of 12.5—23.2 wt%. Obviously, the ore-forming

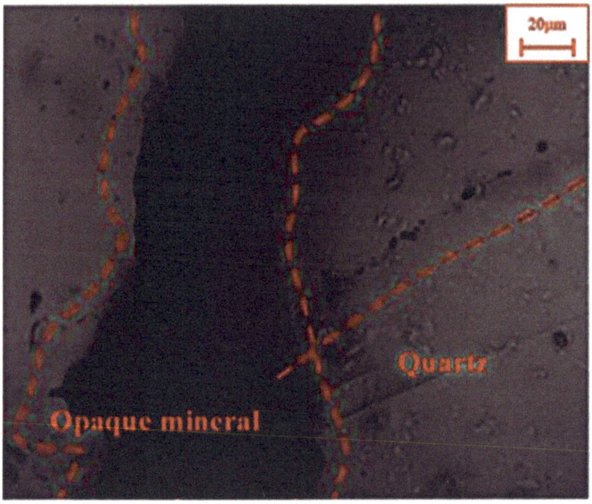

FIGURE 5.12 Contact relations between chalcopyrite and quartz: ore-containing fluid intruding into quartz cracks or dissolving the quartz boundary.

fluids had the properties of hot brine. There were high contents of Na^+ (or K^+), Cl^-, or SO_4^{2-}, as well as aqueous solution of heavy metal components in the hot brine. The overall ratio of SO_4^{2-}/Cl^- (≥ 0.14) showed that it belonged to the chemistry type of sulfuric acid–sodium chloride water. Characterized by high total salinity and SO_4^{2-}/Cl^- ratio, the hot brine was a shallow formation water in genetic classification. Based on the geochemical environment (e.g., neutral to alkaline weak reduction environment and anaerobic bacteria development), metal sulfide of copper was deposited through the combination of Cu and S, forming the structure of a layered copper deposit.

In the associated sheets, the contact relations are observed between opaque mineral chalcopyrite and transparent mineral quartz (see Fig. 5.12). The chalcopyrite may come into contact with quartz (with eroded edge), or intrude on the quartz fissures. This indicates that ore-containing fluid inclusions intrude on quartz cracks or dissolve the quartz during the formation of quartz. Observing the contact boundary, we find that some slit-shaped fluid inclusions in quartz cut through the quartz particles extended into the chalcopyrite boundary. Therefore, fluids in such inclusions are related to mineralization.

After the components of fluid inclusions in chalcopyrite and its associated mineral quartz and calcite were released into the aqueous solution during the grinding process, we determined the concentrations of the released Cu and Fe as well as Cl^- and SO_4^{2-} by ICP-MS and IC, respectively (see Tables 5.10–5.12). The results indicated that the total concentrations of Cu and Fe in chalcopyrite aqueous solution increased with increasing grinding time. Within 6–14 min, C_{CuT} increased from 0.31×10^{-6} to 1.32×10^{-6} mol/L, while C_{FeT} increased from 0.49×10^{-6} to 1.47×10^{-6} mol/L, a significant increase in values.

TABLE 5.10 Concentrations of Cu, Fe, Cl^-, and SO_4^{2-} released from fluid inclusions in chalcopyrite.

	Concentration ($\times 10^{-6}$ mol/L)			
Grinding time (min)	Cu	Fe	Cl^-	SO_4^{2-}
6	0.31	0.49	0.99	10.53
8	0.49	0.62	1.46	15.72
10	0.55	0.74	3.27	22.10
12	1.01	1.20	5.86	27.28
14	1.32	1.47	8.34	30.09

TABLE 5.11 Cu, Fe, Cl^-, and SO_4^{2-} concentration released from fluid inclusions in the associated quartz.

	Concentration ($\times 10^{-6}$ mol/L)			
Grinding time (min)	Cu	Fe	Cl^-	SO_4^{2-}
6	0.11	0.08	2.25	11.52
8	0.31	0.19	2.07	15.99
10	0.29	0.21	2.09	19.10
12	0.37	0.20	3.13	19.90
14	0.40	0.33	5.98	26.01

TABLE 5.12 Concentrations of Cu, Fe, Cl^-, and SO_4^{2-} released from fluid inclusions in the associated calcite.

	Concentration ($\times 10^{-6}$ mol/L)			
Grinding time (min)	Cu	Fe	Cl^-	SO_4^{2-}
6	0.01	0.11	3.78	5.23
8	0.06	0.27	5.91	6.91
10	0.13	0.20	6.45	13.05
12	0.12	0.31	6.54	15.82
14	0.19	0.37	7.27	19.99

The fineness reduced with increased grinding time, so more inclusions were opened into the solution, thus leading to rising concentrations of the components released into the solution.

A small amount of Cu and Fe was detected in the associated quartz and calcite in chalcopyrite, which was related to the formation mechanism of fluid inclusions during ore forming and diagenesis. The nature of contact between the chalcopyrite and the quartz was illustrated in the IR–UV microscopic observation of contact relations between chalcopyrite and the association minerals. Analysis has shown that the concentrations of Cu and Fe in the associated minerals are lower than those in chalcopyrite. The values do not change significantly with the grinding time, because of the uniform distribution of the fluid inclusions in quartz and calcite as well as Cu and Fe elements. With the increasing intensity of grinding, the concentrations of Cl^- and SO_4^{2-} in aqueous solutions of chalcopyrite and its associated quartz and calcite tend to increase with increasing grinding time. ICP-MS values of Cl^- and SO_4^{2-} concentrations in blank water are 7.6×10^{-10} and 20.4×10^{-10} mol/L, respectively, with 3 orders of magnitude difference from those in the solution, indicating that the test and analysis systems have little effect on the results. In addition, it indicates that under the same test conditions, the main source of the ions is the component release of the inclusions rather than self-dissolution of chalcopyrite.

5.4.2 Component release of fluid inclusions in chalcopyrite, sphalerite, and associated minerals

Cu–Zn separation is a worldwide problem in the field of mineral processing. The separation of Cu–Zn sulfide mineral has been selected as a research topic and difficult point by mineralizers. Cu–Zn minerals have small disseminated grain sizes. Some of them have intermetallic inclusions, and some have a great variety of copper sulfide minerals with different degrees of floatability. In addition, some contain lots of secondary copper sulfide minerals and soluble heavy metal salts, which activate and promote the floatability of sphalerite, which is similar to or even higher than that of copper sulfide minerals in solution. These factors bring great difficulties in the separation of copper and zinc. One of the main reasons for this is that sphalerite is activated by the dissolved copper ions after secondary Cu minerals are dissolved in the pulp, and as a result the sphalerite will have the same properties as chalcopyrite.

In terms of flotation separation methods, Clarke believes that different sulfide minerals in the ore can be selectively separated by flotation. Selective suppression of Zn and flotation of Cu is achieved by adding some special components such as inhibitors, polymers, and high-efficiency collectors of Cu mineral in the flotation ore pulp. It is also possible to adjust the pH and Eh of the solution to achieve separation. Such separation method is mainly based on natural reaction characteristics of sulfide minerals, such as surface oxidation and adsorption of the reaction products. The most important influencing factors are the dissolution of minerals, rebuilding the metals, sulfides, species, properties, and charges in the solution. The control of these lysates is important for separation. Only the factors of mineral dissolution and grain size have been mentioned in previous studies, neglecting the release of mineral fluid inclusions' components. The influence of the released fluid inclusion components is also applicable to the separation of polymetallic sulfide minerals, such as Cu—S separation and Cu—Pb separation.

The selected chalcopyrite and sphalerite and their associated quartz were ground in mineral sheets for the study of morphology and component release of inclusions. Fig. 5.13 shows the sample sheets observed under the microscope. The transparent mineral (quartz), translucent mineral (sphalerite), and opaque mineral (chalcopyrite) coexist in the sheet. Visible light and infrared light sources were used to analyze the transparent mineral, translucent mineral, and opaque mineral, respectively. Fig. 5.14 shows the photomicrograph of the chalcopyrite, sphalerite, and quartz under reflected light after treatment of white balance.

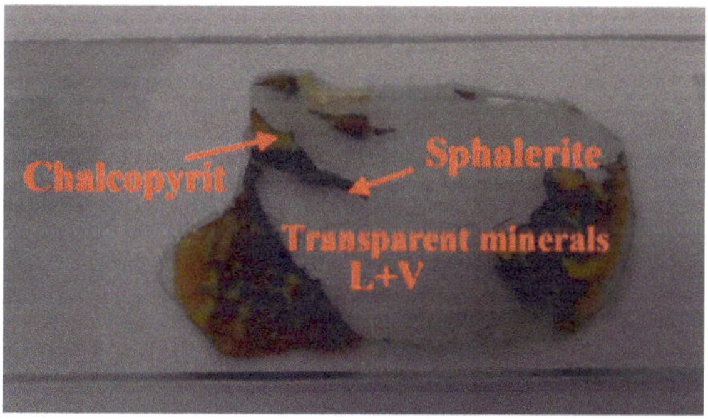

FIGURE 5.13 Fluid inclusion sheets of chalcopyrite, sphalerite, and transparent minerals. *L*, liquid phase; *V*, vapor phase.

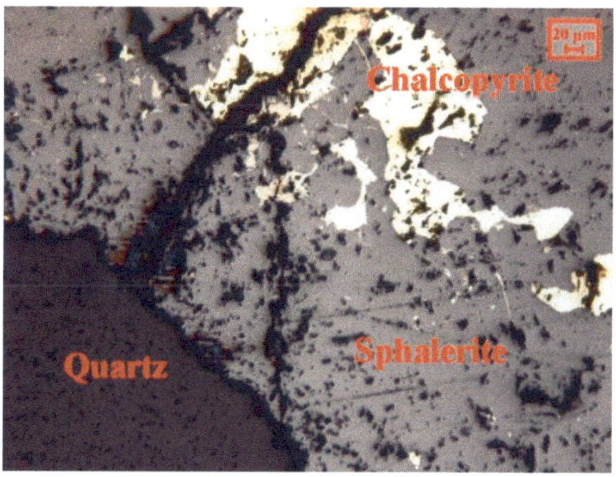

FIGURE 5.14 Micrograph of chalcopyrite, sphalerite, and associated quartz under reflected light.

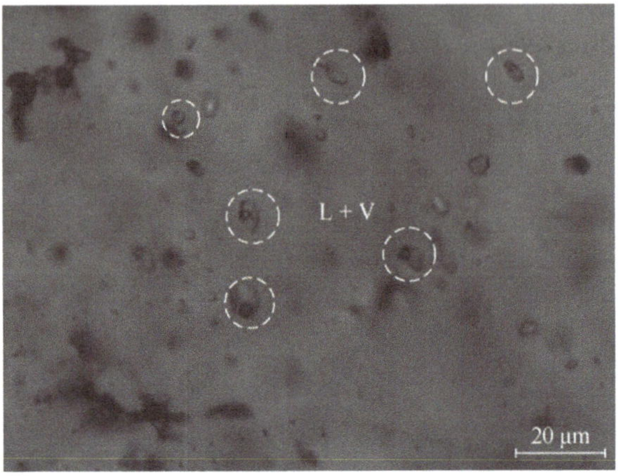

FIGURE 5.15 Fluid inclusion photomicrograph of transparent mineral. *L*, liquid phase; *V*, vapor phase.

Fig. 5.15 shows that there are a large number of clear fluid inclusions in transparent gangue minerals other than those of chalcopyrite and sphalerite. Most of them are distributed in flake form, while a small amount of them are distributed as cracks. By observing the contact characteristics of the cracks, we can find the fluid distribution and the trend at the different mineralization stages. The fluid inclusions vary in size from a few to tens of micrometers. The shapes are regular, circular, elliptical, and even negative crystal, with some irregular shapes. The inclusion types include liquid-rich, gas–liquid fluid inclusions, with a small amount of pure liquid-phase inclusions. The content of the inclusions is usually high.

A large number of fluid inclusions have also been found in the sphalerites associated with chalcopyrite (see Fig. 5.16), mainly distributed along the cracks or in the form of flakes or along mineral boundaries. The contact

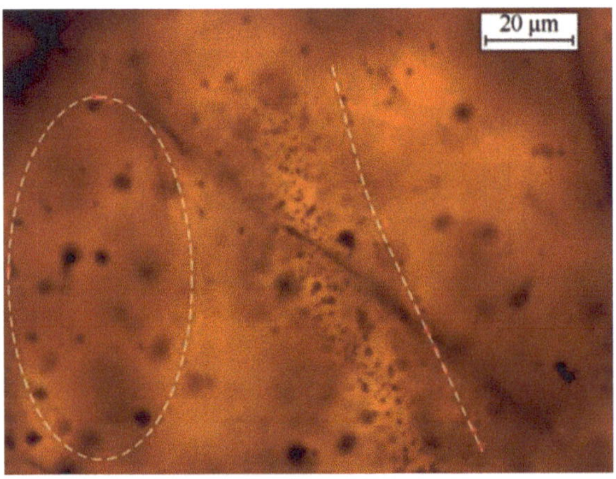

FIGURE 5.16 Micrograph of fluid inclusions distributed along fractures and schistose in sphalerite.

relations (the dashed line) and characteristics can be observed in the fluid inclusions distributed along the cracks. At room temperature, there are many fluid inclusions in the sphalerite, appearing as black pits (the dashed oval). The sizes of the fluid inclusions are small—a few micrometers—and the shapes are generally round and irregular. Under the infrared microscope, many fluid inclusions in chalcopyrite are distributed in schistose forms, with poor imaging effect. The resolvable fluid inclusions vary in size.

After the contents of the fluid inclusions in chalcopyrite, sphalerite, and the associated mineral quartz were released into the slurry solution during the grinding process, we assayed the concentrations of the released Cu and Fe, as well as Cl^- and SO_4^{2-}, by ICP-MS and IC, respectively (see Tables 5.13–5.15).

TABLE 5.13 Concentrations of Cu, Fe, Cl^-, and SO_4^{2-} released from the fluid inclusions in the associated chalcopyrite.

	Concentration ($\times 10^{-7}$ mol/L)			
Grinding time (min)	Cu	Fe	Cl^-	SO_4^{2-}
6	0.09	0.33	0.92	5.12
8	0.11	0.51	1.31	7.91
10	0.39	0.56	2.98	10.00
12	0.78	0.78	4.20	13.14
14	1.02	0.97	6.77	18.03

TABLE 5.14 Concentrations of Cu, Fe, Cl^-, and SO_4^{2-} released from the fluid inclusions in the associated sphalerite.

	Concentration ($\times 10^{-6}$ mol/L)			
Grinding time (min)	Cu	Fe	Cl^-	SO_4^{2-}
6	0.02	0.25	0.77	3.18
8	0.09	0.42	1.02	5.72
10	0.16	0.50	1.97	9.10
12	0.49	0.71	2.19	11.81
14	0.62	0.83	4.12	13.27

TABLE 5.15 Concentrations of Cu, Fe, Cl^-, and SO_4^{2-} released from the fluid inclusions in the associated quartz.

	Concentration ($\times 10^{-6}$ mol/L)			
Grinding time (min)	Cu	Fe	Cl^-	SO_4^{2-}
6	0.03	0.19	1.11	2.99
8	0.10	0.17	0.73	4.73
10	0.21	0.43	1.97	6.36
12	0.29	0.62	2.30	9.16
14	0.44	0.91	3.18	11.00

The above tables show the concentrations of and changes in elements or ions in the solutions of three associated mineral crystals at different grinding times. In the high-purity deionized aqueous washing solution after grinding, there is a magnitude of 10^{-7} or 10^{-6} mol/L Cu, Fe, Cl^-, and SO_4^{2-}. Especially with the increase in grinding fineness, components in the solution gradually increased with more opened inclusions and rise in grinding time. In chalcopyrite aqueous solution, within the grinding time of 6–14 min, C_{CuT} increased from 0.09×10^{-6} to 1.02×10^{-6} mol/L, C_{FeT} from 0.33×10^{-6} to 0.97×10^{-6} mol/L, Cl^- from 0.92×10^{-6} to 6.77×10^{-6} mol/L, and SO_4^{2-} from 5.12×10^{-6} to 18.03×10^{-6} mol/L.

The total concentration of these ions in sphalerite aqueous solution also increased as the grinding time increased. Within grinding time of 6–14 min, C_{CuT} increased from 0.02×10^{-6} to 0.62×10^{-6} mol/L, C_{FeT} from 0.25×10^{-6} to 0.83×10^{-6} mol/L, Cl^- from 0.77×10^{-6} to 4.12×10^{-6} mol/L, and SO_4^{2-} from 3.18×10^{-6} to 13.27×10^{-6} mol/L. Cu, Fe, Cl^-, and SO_4^{2-} were also detected in the chalcopyrite-associated quartz, indicating that ore-forming fluids were immersed in the quartz crystal during the ore-forming process, with mutual penetration. This result is consistent with the microscopic imaging of quartz inclusions in the associated minerals.

Given that the grinding process was involved in the steps of the component release of fluid inclusions, it is implied that metal ions from the grinding medium could also enter the solution. Thus, it is necessary to consider the effects of the grinding medium on the concentration of the metal ions. In this study of fluid inclusions, a blank, or control, was also prepared, with the same ball mill operating parameters and test operation process. The minerals were replaced with a blank sample of deionized water, to investigate the contributions of deionized water and the grinding medium to the contents of Cu, Fe, Cl^-, and SO_4^{2-}. In other words, we used the same small ball mill (MM400, Retsch, Germany), a 50-mL tank containing a stainless steel ball with a diameter of 15 mm. The frequency was set to 900 min^{-1}, with grinding time of 14 min. At the same time, the same proportion of deionized water was added. The contents of Cu, Fe, Cl^-, and SO_4^{2-} were 0.99×10^{-10}, 9.15×10^{-10}, 7.60×10^{-10}, and 20.40×10^{-10} mol/L, respectively. The results showed that the content of Fe was higher than those of the other elements. The contributions of the grinding medium and deionized water to the contents of these elements had approximately 2 orders of magnitude difference from the amounts released from fluid inclusions. Therefore, the concentration had little effect on the test values of the fluid inclusion components, which further demonstrated that the contribution of the German ultraclean ball mill to the solution components could be neglected.

5.5 Morphology and component release of fluid inclusions in sphalerite and associated quartz

5.5.1 Morphology and types of fluid inclusions in sphalerite and associated quartz

The associated quartz of sphalerite was studied to determine whether fluid inclusions in the closely associated gangue minerals (quartz) contained the components of the main mineral (Zn ions). Large sphalerite and quartz crystals were cut into thin sheets with a thickness of 0.1–0.3 mm (see Fig. 5.17). Fig. 5.17 shows that the sphalerite sheets are translucent, while quartz sheets have better transparency.

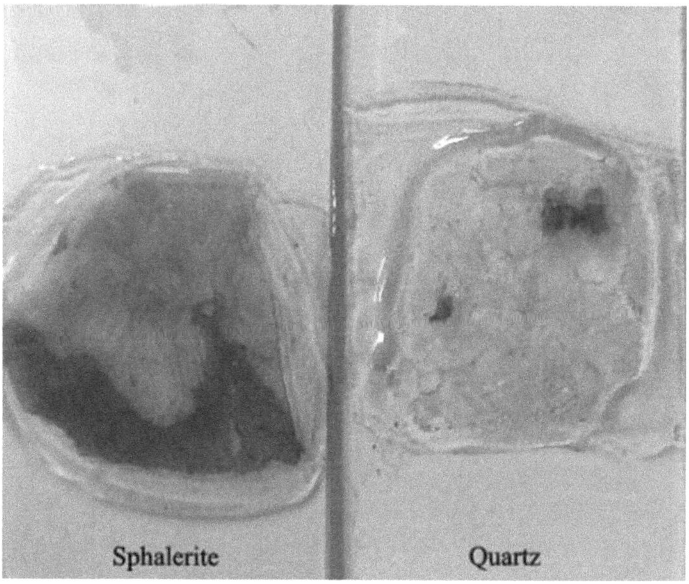

FIGURE 5.17 Thin sheets of natural sphalerite and its associated quartz.

Under infrared light, microscopic imaging was performed with a BX51 microscope. Figs. 5.18 and 5.19 show the morphologies and types of fluid inclusions in the sphalerite and the closely associated quartz sheets, respectively.

Fig. 5.18 shows that the types of fluid inclusions in sphalerite include pure gaseous phase (see Fig. 5.18B), pure liquid phase (see Fig. 5.18A), gaseous–liquid phase (see Fig. 5.18C), gas-rich phase (see Fig. 5.18D), and liquid-rich phase (see Fig. 5.18E). Gas-rich phase and liquid-rich phase refer to fluid inclusions containing over 50% gaseous-phase or liquid-phase components in the inclusion body. The sizes of inclusions in sphalerite range from a few to tens of micrometers, as shown in the group of microinclusions in Fig. 5.18F. Some are even smaller, and difficult to observe under the microscope. Findings have shown that there are a large number of inclusions in sphalerite. The abundance of fluid inclusions in sphalerite sheets is roughly estimated at 7%–10%.

Fig. 5.19 shows that fluid inclusions in the associated quartz of sphalerite are similar to those found in sphalerite. The sizes range from a few to hundreds of micrometers. The abundance of fluid inclusions in quartz under the microscope was roughly estimated at 15%–20%. HRXMT was performed for fluid inclusions in a single quartz particle of 500 μm. The image shows that there are a large number of heterogeneous irregular objects of different sizes in the single quartz particle (see Fig. 5.20), which are the fluid inclusions contained in quartz. The area containing the fluid inclusions presents a darker color, which is a result of the different contributions of the areas containing fluid inclusions to the X-ray signal.

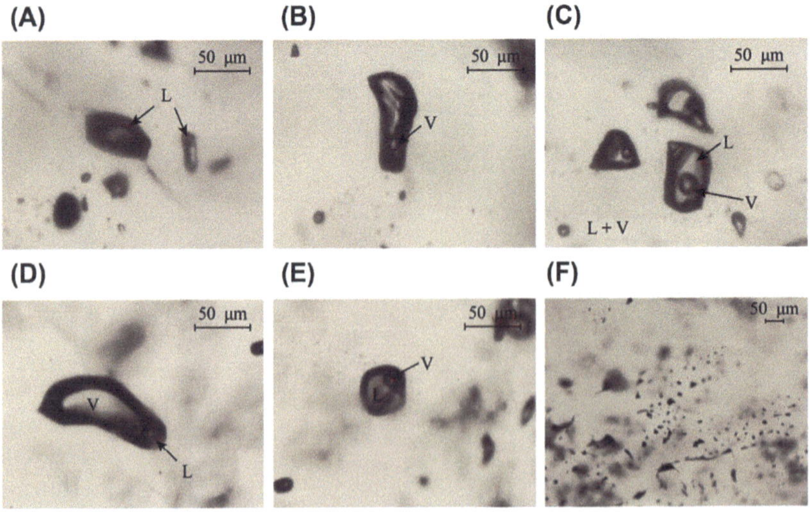

FIGURE 5.18 Morphologies and types of fluid inclusions in sphalerite. (A) Pure liquid inclusion (*L*). (B) Pure gaseous inclusion (*V*). (C) Liquid–gas inclusion (*L* + *V*). (D) Gas-rich phase inclusion. (E) Liquid-rich phase inclusion. (F) Group of microinclusions.

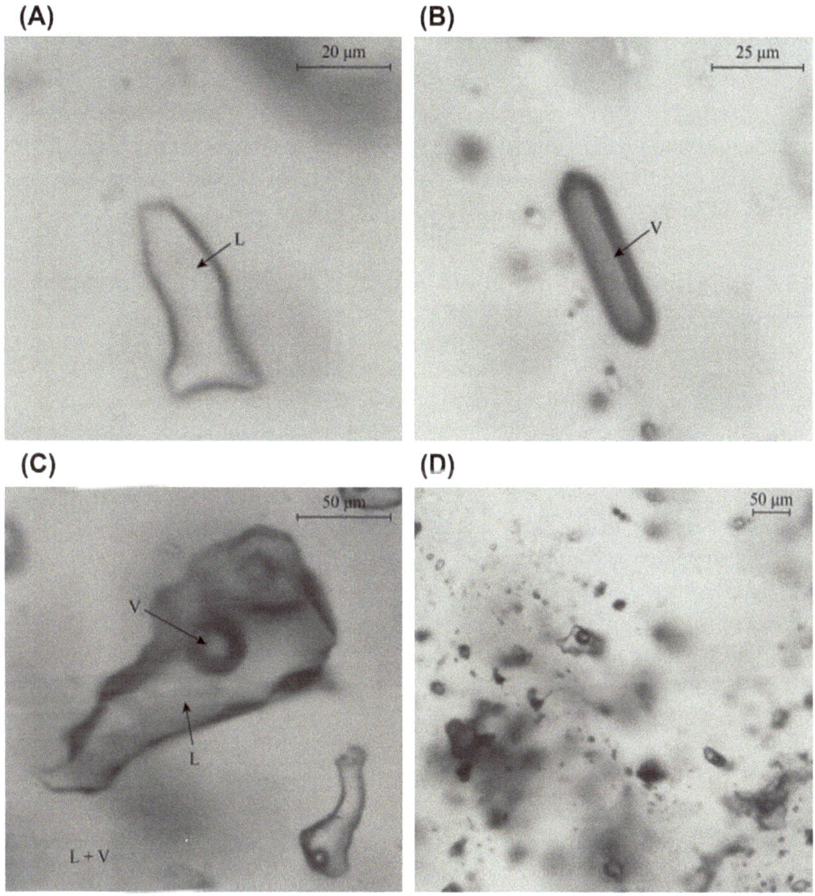

FIGURE 5.19 Morphologies and types of fluid inclusions in quartz. (A) Pure liquid inclusion (*L*). (B) Pure gaseous inclusion (*V*). (C) Liquid–gas inclusion ($L + V$). (D) Group of microinclusions.

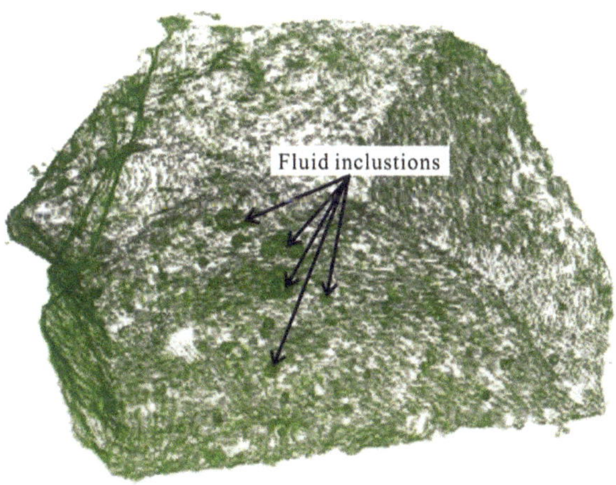

FIGURE 5.20 High-resolution X-ray microfault 3D imaging image of a single quartz particle of 500 µm.

In conclusion, a large number of fluid inclusions are contained in sphalerite and its associated gangue quartz. These fluid inclusions are partially destroyed during the grinding process, resulting in the release of various ions into the slurry solution.

5.5.2 Freezing point and salinity value of the fluid inclusions in sphalerite

The salinity in a single fluid inclusion can be calculated by formula. First of all, the freezing point is obtained by microscopic temperature measurement of a single fluid inclusion. Then, the freezing point is substituted into the empirical formula to obtain the salinity of the inclusion. Therefore, the appearance of Cl^- in the aqueous solution of ground sphalerite is an important basis for the component release of the fluid inclusions during the grinding process.

The transparency of the mineral will change with an increase or decrease in the temperature of sphalerite under infrared light. In this case, general microscopic temperature measurement technology cannot meet the requirements. Hence, in this work, cyclic temperature measurement technology was adopted to determine the freezing point of the inclusions in sphalerite. The test was completed on a heating and freezing stage (THMS-600, Linkam, UK), using the following procedure.

First of all, the size, shape, and location of the bubbles in the fluid inclusions were observed at 20°C. Then, the bubbles in the fluid inclusions were smashed or made to disappear by lowering the temperature to the supercooled state (−60°C). Upon heating to −25°C, a small bubble began to appear in the inclusions; when heated to −20.7°C, the bubble slightly enlarged and continued as the temperature rose from −18.3°C until it approached its state at room temperature. We observed the changes in bubble size as the temperature was gradually raised at a rate of 0.2°C/min until room temperature was reached. To confirm whether the ice in the inclusion had completely melted, we rapidly lowered the temperature to a certain value slightly lower than the recorded temperature. If the size of the bubble suddenly decreased, it would indicate that the ice did not completely dissolve. The process was repeated again and again by continually raising the temperature until the bubble was close to the state at room temperature. The point at which the bubble size did not change after rapid cooling or lowering of the temperature was recorded as the freezing point. Thus, the temperature became the freezing point of the fluid inclusions when the bubble was close to the state at room temperature. After obtaining the freezing point of the fluid inclusions, the salinity of the fluid inclusions was calculated by:

$$S = 1.78A - 0.0442A^2 + 0.000557A^3 \tag{5.1}$$

where A is the absolute freezing point value (°C) of the fluid inclusion and S, the salinity (wt%, NaCl). The previous analysis found that there are a large

TABLE 5.16 Freezing point and salinity of the sphalerite fluid inclusions.

No.	Type	Gas/liquid ratio (%)	Mean FP (°C)	Mean salinity (%)
1	Gas–liquid inclusion	5–40	−9.20	13.07
2	Gas–liquid inclusion	8–30	−10.70	14.67
3	Gas–liquid inclusion	25–30	−8.20	11.93
4	Gas–liquid inclusion	5–33	−8.18	11.91
5	Gas–liquid inclusion	5–40	−10.69	14.66
6	Gas–liquid inclusion	10–30	−12.05	16.00
Average	—	—	—	13.71

FP, freezing point.

number of fluid inclusions in sphalerite. Therefore, it is impossible to determine the freezing point and salinity of each inclusion by microscopic temperature measurement. The partition sampling method was adopted to select six well-formed gaseous–liquid-phase inclusions from the sheets under infrared microscopy for measurement. Table 5.16 shows the measured freezing points and the salinity values calculated by Eq. (5.1). It indicates that the gas/liquid ratio of the gaseous–liquid-phase fluid inclusions in sphalerite ranges from 5% to 40%. There are great differences in the gas/liquid ratios and freezing points among the various inclusions. However, the variation in the salinity is small, ranging from 11.91% to 16%. The theoretical average salinity is 13.71%; thus sphalerite inclusions are rich in sodium chloride.

5.5.3 SEM/EDS analysis of the fluid inclusions on sphalerite surface

Some micrometer-sized or even smaller damaged areas are left on the surface after the fluid inclusions of sphalerite surface have been destroyed. Due to adsorption on the surface, some components of fluid inclusions are retained in these damaged areas of the inclusions. Thus, the composition of fluid inclusions can be indirectly obtained by studying the components of fluid inclusions on the surface, further confirming the existence of fluid inclusions.

SEM and EDS were used to research the geometry and chemical composition of the destroyed areas in the fluid inclusions of the sphalerite surface. Fig. 5.21 shows the SEM morphology and EDS spectrum of the damaged areas of sphalerite inclusions. Fig. 5.21F shows the morphology of the sphalerite surface. Fig. 5.21A–E shows the EDS spectrum results corresponding to the five points in Fig. 5.21F.

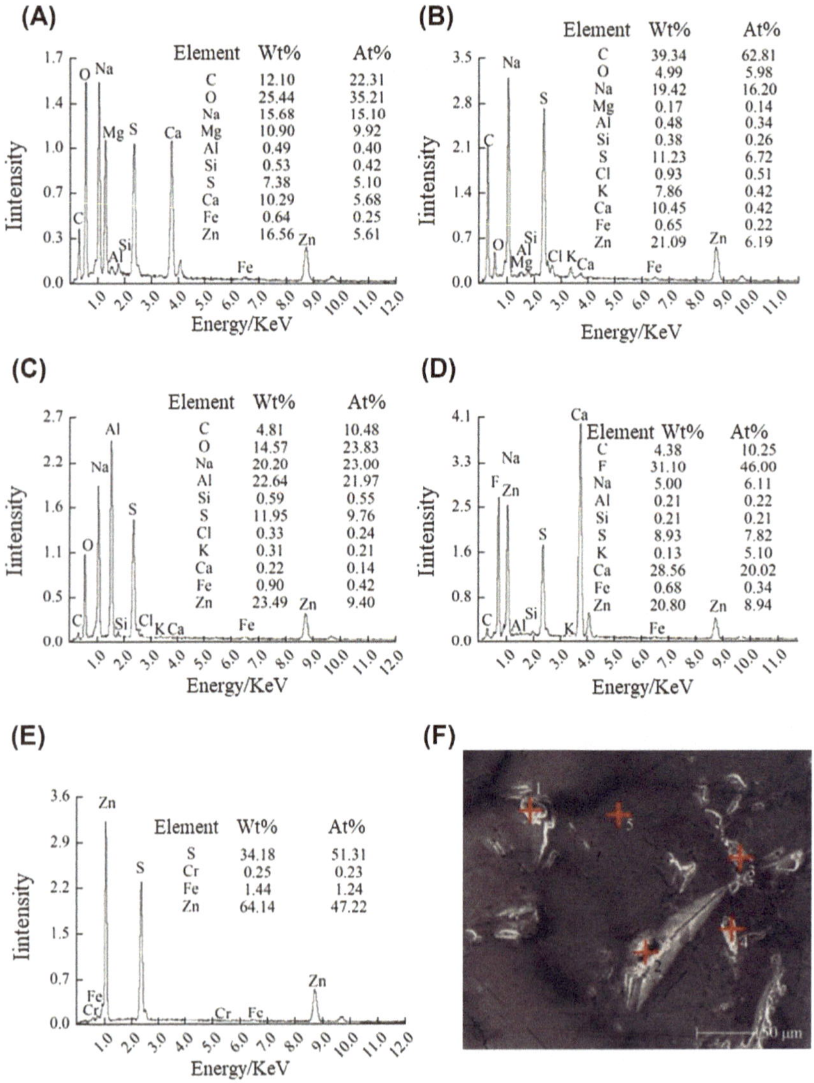

FIGURE 5.21 Scanning electron microscopy and energy-dispersive X-ray spectroscopy results of inclusions in the destroyed areas of sphalerite. (A–E) Positions *1–5*,respectively, shown in (F). (F) Surface morphology.

Fig. 5.21F indicates that there are many concave damaged areas of different shapes and sizes on the sphalerite surface. The damaged areas originally consisted of fluid inclusions. During grinding, the fluid inclusions were destroyed, leaving the residual micrometer-sized damaged areas. The flat areas in Fig. 5.21F mean that there are no fluid inclusions in these areas, or the inclusions in these areas have not been destroyed. This work attempted to prove that these damaged areas were the residual areas left after destruction of the fluid inclusions rather than the scratches left by sheet preparation. EDS energy analysis was performed on the damaged areas and on the flat areas of Fig. 5.21F. To rule out contingency of the experiment, we selected the points of four damaged areas (points 1–4) and the point of one flat area (point 5) for analysis. Fig. 5.21E shows that the flat area on the sphalerite surface mainly comprises Zn and S, with contents of 64.14% and 34.18%, respectively. This is consistent with the results obtained from previous chemical analysis of pure mineral. The atomic concentration ratio of Zn/S is 0.92, which is close to the theoretical atomic ratio of sphalerite of 1:1. In addition, we detected small amounts of Fe and Cr, which are common impurity elements in natural sphalerite. Therefore, the flat area is mostly composed of the sphalerite components.

However, it was observed that the composition of the four damaged areas was significantly different from that of the flat area. Fig. 5.21A–D shows that large amounts of alkali metal and alkaline earth-metal elements exist in the damaged areas, including K, Ca, Na, and Mg. These elements are common components of fluid inclusions. The alkali metal and alkaline-earth metal elements can be derived only from the component release of fluid inclusions for the following reasons. After the fluid inclusions of sphalerite have been destroyed, the components are released to form damaged recessed areas. Some of the released components are adsorbed and remain on the newly formed damaged surface. In addition, the contents of Zn and S in the damaged areas deviate significantly from the theoretical values. If the damaged areas of Fig. 5.21F are scratches caused by the grinding process, the composition should be the same as or similar to that of flat areas. EDS analysis clearly indicates that the elemental composition of the damaged areas is significantly different from that of the sphalerite surface. Therefore, the damaged area is retained after the destruction of fluid inclusions on the sphalerite surface. In some depressed areas, Cl^-, which is the most common component of mineral fluid inclusions, is particularly detected (see Fig. 5.21B and C). The presence of Cl^- provides strong evidence for the view of this work.

5.5.4 Component release of fluid inclusions in sphalerite and quartz

Existing studies have shown that fluid inclusions in sphalerite mainly contain the cations Mg^{2+}, Ca^{2+}, Na^+, K^+, and Li^+ and heavy metal ions, and the anions Cl^-, F^-, SO_4^{2-}, and HCO_3^{2-}. Direct measurement methods have been

TABLE 5.17 Concentrations of Zn and Cl^- released from the sphalerite fluid inclusions.

	Concentration ($\times 10^{-6}$ mol/L)	
Grinding time (min)	Zn	Cl^-
2	6.24	39.72
4	6.88	20.28
6	7.12	13.52
8	8.79	10.99
10	18.35	8.92

established to study the composition and content of these ions. For example, the laser plasma ablation method (LA-ICP-MS), which has been widely used in the direct determination of the content of the various elements in inclusions, has achieved good results. However, the direct determination of Zn content is still a difficult problem, because the method cannot eliminate the contribution of the sphalerite inclusion wall (Zn in zinc sulfide) to the content of Zn in the inclusions. Therefore, the indirect measurement method was adopted in this work, namely using the grinding–washing method, to determine the Zn content in sphalerite fluid inclusions. Table 5.17 shows the results of ICP-MS/AES and IC detection of Zn and Cl^- in sphalerite fluid inclusions.

Table 5.17 shows that C_{ZnT} in the sphalerite wash water significantly increases with the increase in grinding time. As the grinding time increases from 2 to 10 min, the C_{ZnT} in the corresponding wash water increases from 6.24×10^{-6} to 18.35×10^{-6} mol/L. The reason is that the number of inclusions destroyed increases as the fineness of the particles increases during grinding, leading to increased amount of Zn released into the solution. A large amount of chloride ions also appeared in the wash water after the grinding of sphalerite, which can only be released from the fluid inclusions in sphalerite, as there is no chloride in the ultrapure deionized water. However, the concentration of the chloride ions in the wash water decreased significantly with increasing grinding time. This abnormal phenomenon may have the following cause. Skou et al. found that zinc ions react with chlorine in aqueous solutions containing chlorine as follows:

$$Zn^{2+} + nCl^- = [ZnCl_n]^{2-n} \quad (n = 1-4) \tag{5.2}$$

Eq. (5.2) indicates that zinc–chloride complexes are generated by the combination of Zn^{2+} in aqueous solution and chloride ions of different coordination modes. Zn can exist in the form of Zn^{2+}, $ZnCl^+$, $ZnCl_2$, $ZnCl_3^-$, and $ZnCl_4^{2-}$. Thus, the concentration of chloride ions in the wash water will decrease with the increase of Zn, because the chloride ions form complexes with zinc ions. Since IC can detect only the ionic chlorine and not the chlorides in these complexes, ICP-MS/AES was used to detect the concentrations of all forms of Zn in the solution.

Table 5.17 indicates that there are large amounts of Zn in sphalerite wash water. This work attempted to further confirm that they were derived from the release of sphalerite fluid inclusions and not just from dissolution of the sphalerite. A sphalerite dissolution test was performed. First of all, 2 g of the sphalerite sample from the maximum grinding time (10 min) was taken and placed into the centrifuge tube. Then, 40 mL of deionized water was added for 1 min of ultrasonic washing. After the washing, the tube was placed into the centrifuge for solid–liquid separation. The obtained solid was naturally dried in an argon atmosphere for subsequent dissolution experiments. The process was chosen to remove most of the Zn released by fluid inclusions in the sphalerite during the grinding. However, this method could not completely eliminate the interference of Zn released from the inclusions, because the inclusions were not fully opened, but only cracks were generated in the inclusions during the grinding process. Thus, the rate of release of the inclusion components was slow. The dissolution test was carried out on a magnetic stirrer. Two grams of sphalerite that was naturally dried after the removal of inclusion components and 40 mL of deionized water were placed in a 50-mL glass reactor and stirred at a rate of 1000 rpm. After stirring, the sample was left to dissolve, and was exposed to air for 3 h. Next, the solid–liquid separation was carried out by centrifugation, the supernatant was collected, and the concentration of Zn in the solution was measured by ICP-MS. Table 5.18 shows the experimental results.

Table 5.18 shows that the concentration of Zn in solution after the dissolution of sphalerite in deionized water at a natural pH is only 1.93×10^{-6} mol/L despite the long-term aerobic dissolution. This value is lower than that of Zn in the sphalerite wash water under the same grinding conditions, namely 18.35×10^{-6} mol/L. In addition, according to the theoretical calculation of sphalerite solubility, the theoretical value of Zn dissolved in sphalerite at pH 6.8 is 9.18×10^{-8} mol/L, indicating that the solubility of sphalerite is very small in neutral water. Sphalerite is a hardly soluble sulfide. The dissolution experiments and theoretical calculations of sphalerite were performed after the

TABLE 5.18 Concentration of Zn released from sphalerite dissolution after removing fluid inclusions.

Grinding time (min)	Dissolution time (h)	pH	C_{ZnT} (mol/L)
10	3	6.8	1.93×10^{-6}

TABLE 5.19 Concentrations of Zn and Cl^- released from the quartz fluid inclusion.

	Concentration ($\times 10^{-6}$ mol/L)	
Grinding time (min)	Zn	Cl^-
0.5	0.87	93.80
1	1.12	61.69
2	0.89	74.37
4	0.64	87.32
10	0.88	70.14

complete removal of the fluid inclusions. The results demonstrate that the Zn in the sphalerite wash water of Table 5.17 is derived from the release of sphalerite fluid inclusions rather than the self-dissolution of sphalerite.

During the ore-forming process, gangue minerals also capture the ore-forming fluids of some main minerals due to the closely associated relationship between them, and since quartz is a closely associated mineral of sphalerite, if Zn is detected in the wash water after grinding of quartz, it can further prove indirectly that Zn is contained in the sphalerite fluid inclusion. This experiment further demonstrates that the gangue mineral is the main contributor of metal ions in the slurry solution. Table 5.19 shows the detection results of Zn and Cl^- in quartz.

Table 5.19 shows that there are a large amount of Cl^- and a small amount of Zn in the quartz wash water, indicating that Zn is contained in the fluid inclusions of the gangue minerals associated with sphalerite. It also demonstrates that Zn is contained in the fluid inclusions of the main mineral sphalerite. However, the concentration of Zn released by the quartz inclusions is much lower compared with that released from sphalerite inclusions, consistent with the results predicted by Wilkinson et al. For example, at the same grinding time (10 min), the concentration of Zn in quartz wash water (0.88×10^{-6} mol/L) is 21 times lower than the concentration in sphalerite wash water (18.35×10^{-6} mol/L). The presence of Zn in the quartz is because due to diagenetic fluids being mainly captured by the gangue mineral during mineralization. The amount of the ore-forming fluids captured in the gangue mineral is much smaller compared with the content in the main mineral. In addition, the concentrations of Zn and Cl^- in the quartz wash water are not as evenly distributed as in the sphalerite wash water. Instead, the concentrations present an irregular distribution with grinding time, which may have two reasons. First, quartz is more fragile compared with sphalerite, so it can be fully ground in a short time at the same grinding frequency. Therefore, there is no linear relationship between the Zn released by the inclusions in quartz and the grinding time. Second, the content of Zn is low in fluid inclusions of quartz. Thus, the distribution and content of Zn and Cl^- in each inclusion may be significantly different from those of the main minerals and, therefore, present high randomness, which is related to the opportunity to capture the ore-forming fluids during the quartz mineralization.

In summary, Zn is contained in both the sphalerite and its closely associated gangues, with important significance for its actual beneficiation. The actual mineral ore is mainly composed of gangue minerals, and although the content of the main mineral is very low in the fluid inclusions of the gangue minerals, we cannot completely ignore the influence of the amount of heavy metal ions released by the fluid inclusions of the gangue minerals during mineral processing.

5.6 Morphology and component release of fluid inclusions in galena

5.6.1 Infrared optics and SEM/EDS of fluid inclusions in galena

Fig. 5.22 shows the structure of the fluid inclusions on the surface and superficial layers of galena. According to the results of the infrared optical microscopy, there are a large number of fluid inclusions in galena. Fig. 5.23 shows the morphology of the dissociation surface after opening the inclusions in galena. EDS was used to analyze the suspected destroyed area of the inclusions on the flat and depressed areas of the surface. Fig. 5.24 shows that the elemental composition of these depressed areas is obviously different from that of flat areas. The ratio of Pb to S significantly deviates from that of the stoichiometry of galena, which further proves that the depressed areas remain during the opening of the inclusions. These foreign components confirm that surface adsorption of chemical components occurs on inclusion walls at the end of inclusion destruction.

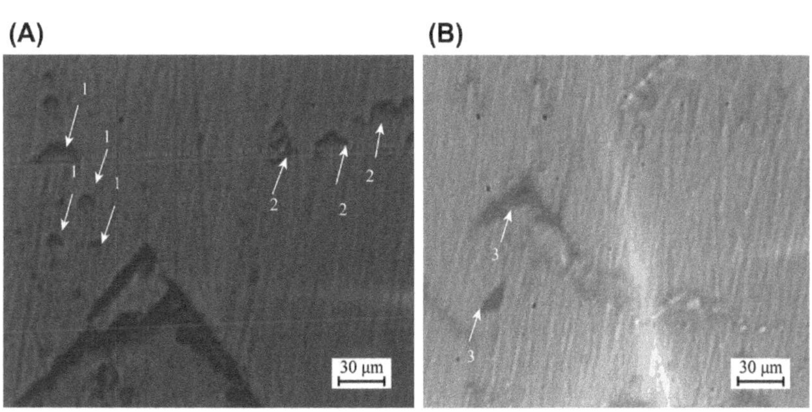

FIGURE 5.22 Infrared micrographs of superficial fluid inclusions in galena : (A) Oval shape inclusions with clumping groups (1 and 2), and (B) strip shape inclusions with isolated position (3).

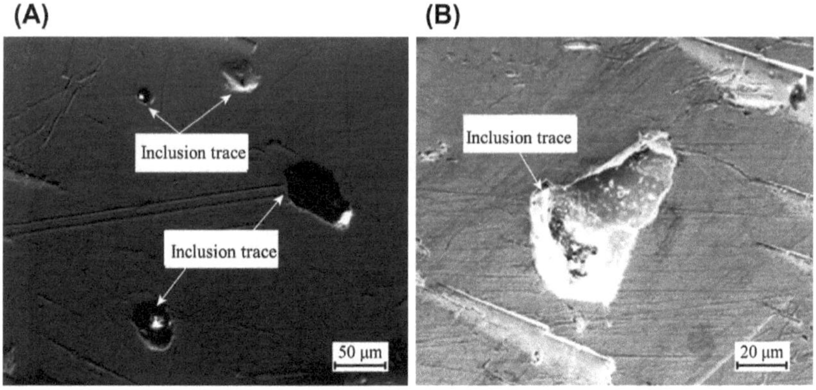

FIGURE 5.23 Scanning electron microscopy of the residuals after breakage of the fluid inclusions in galena : (A) inclusion traces observed in position 1 and (B) inclusion trace observed in position 3.

5.6.2 Component release of inclusions in galena

After crushing of the inclusions in galena, ICP-MS and IC were used to detect the concentrations of Pb and Cl^- in the solution, respectively (see Table 5.20). The results show that lead ions in the aqueous solution significantly increase with the grinding time, because more inclusions in the galena are opened as the grinding time increases, resulting in the release of more Pb into the solution. Similar to the component release of quartz inclusions, in galena, the concentration of Cl^- in the solution significantly decreased with time, and this is because the inclusions were removed in the galena at the same grinding time. The washed samples were placed in deionized water at natural pH for 3 h of dissolution. ICP found that the concentration of Pb was only 1.12×10^{-6} mol/L, much lower than that in the galena wash water under the same grinding conditions (8.25×10^{-6} mol/L). This further confirms that the Pb released from fluid inclusions in galena is the main source of Pb in the aqueous solution, rather than self-dissolution of galena.

5.7 Morphology and component release of fluid inclusions in pyrite

5.7.1 Infrared optical microscopic analysis of fluid inclusions of pyrite in Weixin

Mineral lattices of transparent or translucent minerals, such as quartz, calcite, fluorite, stone salt, garnet, apatite, dolomite, barite, topaz, and sphalerite, can be easily viewed by infrared optical microscopy. However, this method presents a high randomness and difficulty in observing inclusions of opaque minerals such as pyrite, chalcopyrite, and galena, although the structure of the surface and superficial layers can still be observed (see Fig. 5.25).

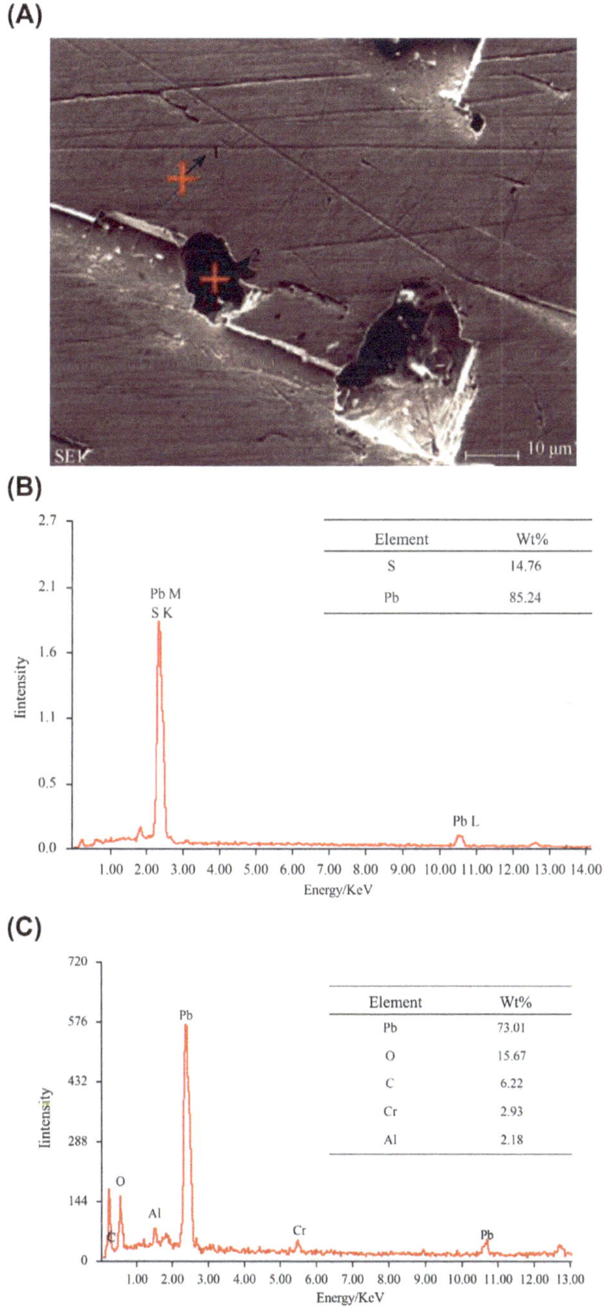

FIGURE 5.24 Energy-dispersive X-ray spectroscopy of the residual depressed areas of fluid inclusions in galena: (A) Surface morphology, (B) and (C) corresponding to the positions of 1 and 2 in (A), respectively.

TABLE 5.20 Concentrations of Pb and Cl^- released from fluid inclusions in galena.

	Concentration ($\times 10^{-6}$ mol/L)	
Grinding time (min)	Pb	Cl^-
4	2.13	90.85
6	3.88	80.54
8	4.32	53.42
10	6.54	45.78
12	8.25	35.43

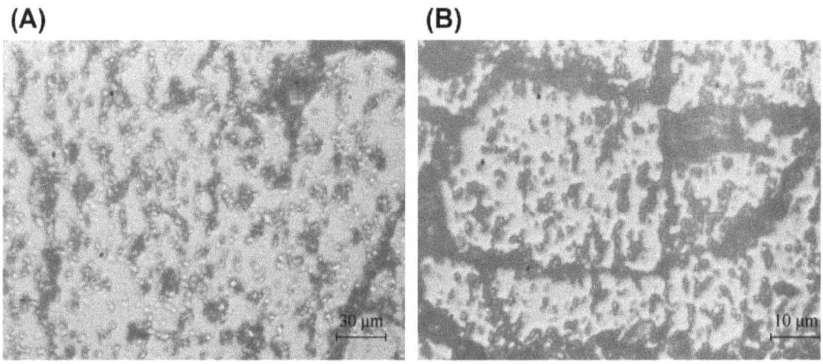

FIGURE 5.25 Infrared optical microscopy images of the surface and superficial layers of pyrite from the Weixin area: (A) image at low magnification (B) image at high magnification.

Fig. 5.25 shows visibly brighter and darker areas, which indicates the compositional differences in these areas. Since the sample is relatively pure, it is inferred that the compositional differences mainly resulted from the substances formed on the inner wall of the inclusions. Such substances are generated after the fluids in the mineral inclusion react with the main mineral. The differences lead to variation in the absorption intensities of the infrared light, thus resulting in brighter and darker areas. Fig. 5.25 shows that these suspected inclusions in the pyrite are isolated or in groups, varying in striped, elliptical, and irregular shapes. The sizes range from a few to tens of micrometers. Since it is difficult to accurately determine the inclusions in opaque mineral by this method alone, additional techniques should be adopted to enhance the viewing of these mineral inclusions.

5.7.2 SEM/EDS detection of the positions of inclusions in pyrite from Weixin

The pyrite sample collected from Weixin was fixed on a scanning electron microscope. Then, a low-power microscope (about 500×) was used to find broken inclusions. Semiquantitative analysis was performed on the target positions using an X-ray energy dispersive spectrometer (working power of 0.5−30 kV).

Fig. 5.26 shows the morphology of the dissociated surface after destruction of the inclusions in pyrite. The SEM surface imaging shows that a certain

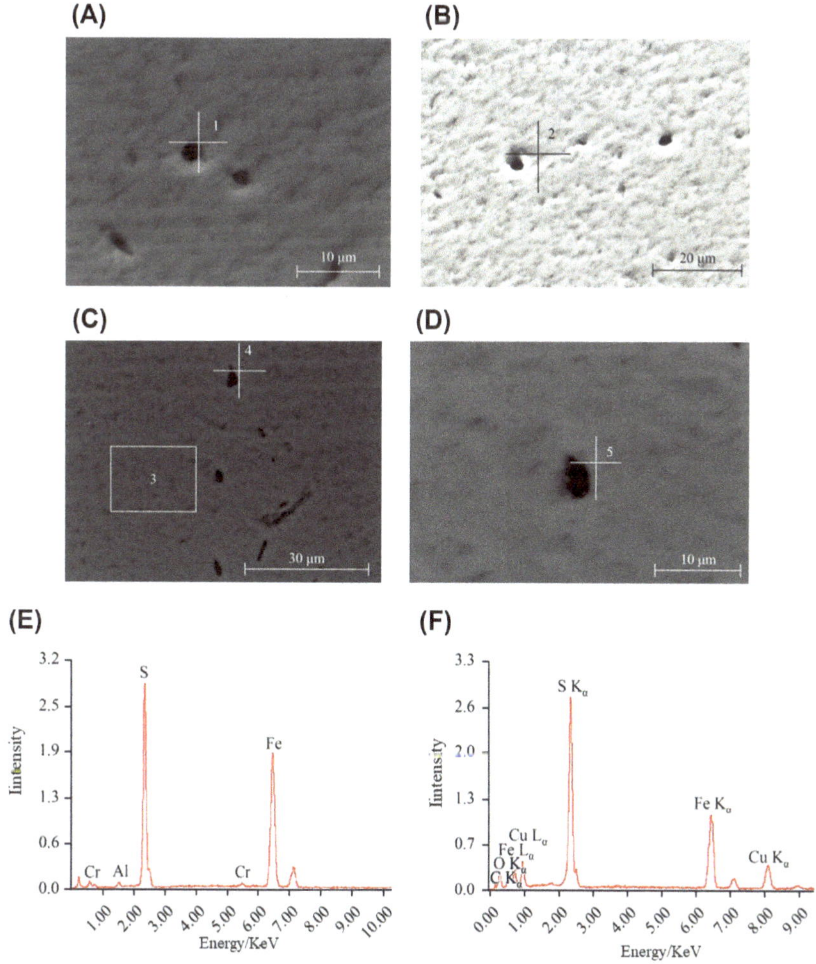

FIGURE 5.26 (A) Secondary electron image of pyrite sheets. (B, C, D) Backscattered electron images. (C, D) The same area with different magnifications. (E−I) Energy-dispersive X-ray spectroscopy results. (E) Position 1, (F) position 2, (G) position 3, (H) position 4, (I) position 5.

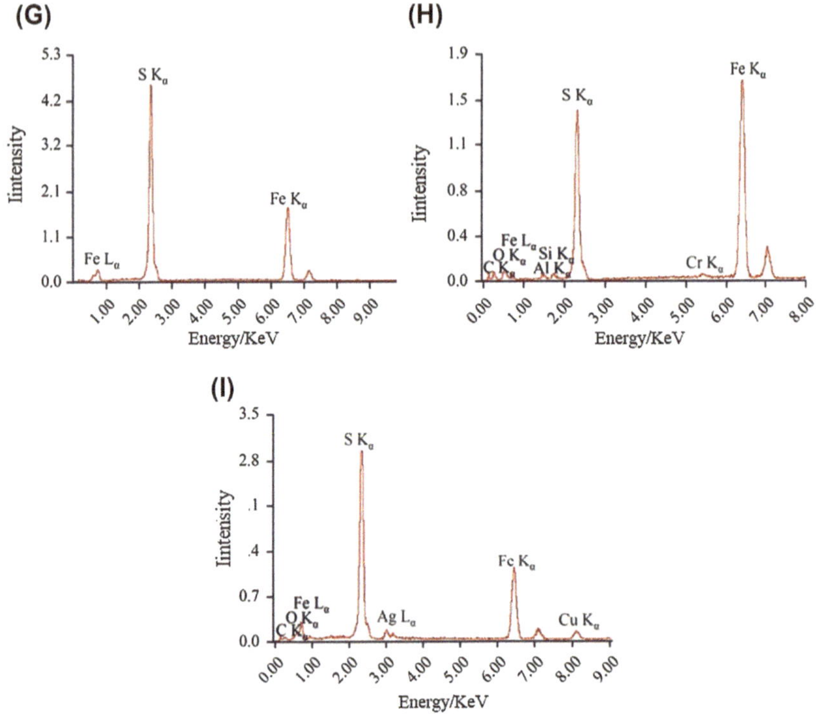

FIGURE 5.26 cont'd

number of dimples occur on the dissociation surface of the pyrite, namely the traces left after opening the inclusions. The structure and morphology of the inclusions are observed by SEM. It revealed that the inclusions have varying sizes and shapes. The sizes range from a few to tens of micrometers. Some of the inclusions are isolated, while some are concentrated. The shapes are striplike, elliptical, round, and irregular. The SEM topography results are consistent with those of the infrared optical microscopy imaging of fluid inclusions on the surface and superficial layers of pyrite. EDS analysis was performed to further prove that the dimples on the dissociation surface were residues left after the destruction of the inclusions, rather than scratches caused by surface grinding.

For comparison, the following positions and regions were analyzed: the depressed positions (1, 2, 4, and 5) and the flat area (3) around the dimples (see "+" and boxes in Fig. 5.26A–D). Fig. 5.26E–I shows the EDS spectrum, and Table 5.21 shows results of the semiquantitative element content. The results indicate that the dimples contain small amounts of Al, Cr, C, O, Ag, and Cu in addition to S and Fe of the pyrite. The homonymous ions (Fe atoms) in positions 1 and 4 are twice the theoretical values, demonstrating that the fluids in the inclusions also carry mineralization elements.

TABLE 5.21 Semiquantitative analysis results of the element concentrations determined by energy-dispersive X-ray spectroscopy.

Scanning position	Element	Wt%	At%
1	Al	1.69	2.66
	S	39.98	52.94
	Cr	1.12	0.91
	Fe	57.22	43.5
2	C	28.55	57.85
	O	0.31	0.47
	S	35.08	26.63
	Fe	23.39	10.2
	Cu	12.67	4.85
3	S	54.73	67.8
	Fe	45.27	32.2
4	C	9.23	26.54
	O	0.2	0.44
	Al	1.92	2.46
	Si	1.38	1.7
	S	32.27	34.78
	Cr	1.01	0.67
	Fe	53.99	33.4
5	C	06.36	19.07
	O	00.18	00.41
	S	48.35	54.33
	Ag	07.91	02.64
	Fe	31.43	20.28
	Cu	05.77	03.27

The electron energy spectrum of position 3 shows only the peaks of S and Fe. Meanwhile, the semiquantitative analysis results show that the concentration ratio of S to Fe is very close to 2:1, which is the chemometry for the theoretical composition of pyrite. Obviously, the pyrite surrounding the dimples does not contain impurity elements. It is a pure pyrite crystal surface, different from the dimples on the sheets. According to the EDS spectrum and semiquantitative results, the depressed areas naturally exist on the surface, rather than being products of the grinding of the surface. The impurity elements of the phase interface components detected by EDS are inherently derived from the residues after release of the fluid inclusion. In other words, these depressed areas are residues from the opening of the inclusions. The various ions are adsorbed on the inclusion wall after volatilization of the gas-phase and liquid-phase inclusions.

5.7.3 High-resolution X-ray topography of microfault pyrite from Weixin

The structure of the fluid inclusion holes in the pyrite particles was scanned, and the scanning morphology is shown in Fig. 5.26. In the low-magnification high-resolution X-ray topography image (see Fig. 5.27A), it is easy to distinguish between the fluid inclusions and the minerals. The circle in Fig. 5.27A is the fluid inclusion hole. The inclusions are isolated from the entire mineral surface. Fig. 5.27B shows the high-magnification image of the circled hole. The diameter of the inclusion hole was measured to be about 25 μm, which is consistent with the aforementioned findings.

Fig. 5.28 shows the CT scanning results of the hole. The straight line in Fig. 5.28A is the scan range; Fig. 5.28B is the scanning result. The results show that the CT value of the hole is significantly higher than that of the outer hole area, indicating that there are other elements inside the hole, which is consistent with the EDS results.

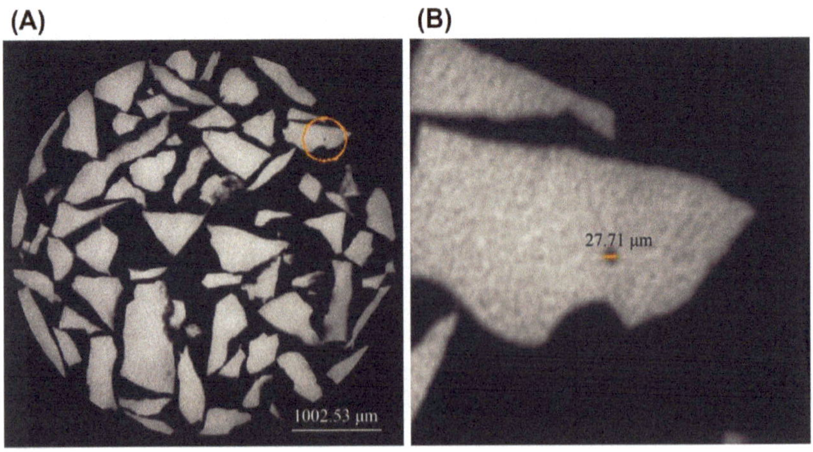

FIGURE 5.27 High-resolution X-ray topography images of pyrite. (A) Low magnification. (B) High magnification.

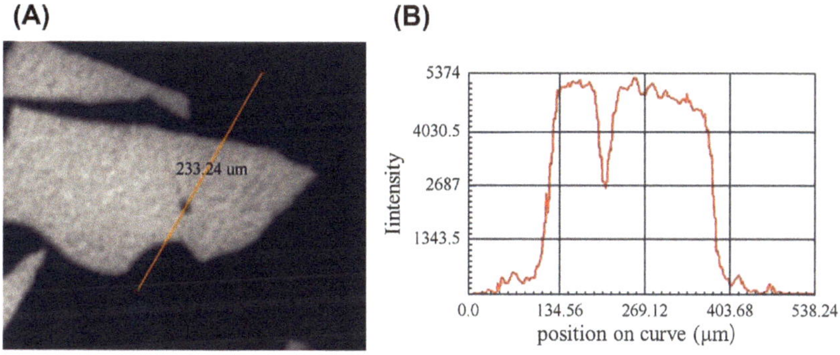

FIGURE 5.28 CT scan of fluid inclusion in pyrite. (A) Scanned area. (B) Scanning result.

5.7.4 Component release of the fluid inclusions of pyrite from Weixin

The cationic components of the fluid inclusions in pyrite contain alkali metal and alkaline-earth metal ions, such as Mg^{2+}, Ca^{2+}, Na^{+}, and K^{+}. The anions include Cl^{-}, SO_4^{2-},and F^{-}, which has been proven in geochemistry. However, there are few geochemical studies on heavy metals and excessive metal ions such as Cu, Pb, Zn, and Fe. ICP-MS and IC were used to detect the concentration of each component released into the solution after crushing of the pyrite. To ensure the scientific agreement of the test, seven parallel tests were carried out to obtain the average values (see Table 5.22).

TABLE 5.22 Concentrations of Cu, Pb, Zn, Fe, Cl^{-}, and SO_4^{2-} in solution released from the macrodefects of pyrite.

	Concentration ($\times 10^{-6}$ mol/L)					
Test No.	Cu	Pb[a]	Zn	Fe	Cl^-	SO_4^{2-}
1	3.81	0.90	0.94	27.82	70.86	87.93
2	3.87	1.73	0.94	31.45	38.57	66.64
3	3.12	0.69	1.37	34.71	56.00	83.79
4	2.54	1.11	2.24	38.29	46.00	61.29
5	2.64	0.50	1.12	31.25	36.57	78.02
6	3.08	0.68	0.81	38.39	82.57	71.72
7	3.95	0.98	1.15	25.76	30.40	81.90
Average	3.29	0.94	1.23	32.52	51.57	75.90

[a] $\times 10^{-8}$ mol/L.

Cu, Pb, and Zn were released from the pyrite inclusions in the aqueous washing solution. Table 5.22 shows that in the seven parallel tests, the average concentrations of Cu, Pb, and Zn were 3.29×10^{-6}, 0.94×10^{-6}, and 1.23×10^{-6} mol/L, respectively. Hydrothermal-mineralized pyrite deposits are often accompanied by interactions with brines in the mineralization process. Therefore, the anion solution after the release of the inclusions will contain anions of Cl^- and SO_4^{2-}. The average concentration values of Cl^- and SO_4^{2-} were 51.57×10^{-6} and 75.90×10^{-6} mol/L, respectively. Table 5.22 demonstrates that there were fluctuations in the test results despite the same experimental conditions and operations being maintained. This indicates that a certain degree of nonuniformity and randomness exists in the distribution of the fluid inclusions and component contents in pyrite crystals.

Furthermore, we attempted to prove that the components in the aqueous solution did not result from self-dissolution of the pyrite mineral. After the complete elimination of the inclusion components, 2 g of pyrite sample was dried and then placed into a 50-mL glass beaker with 40 mL of deionized water. After that, the beaker was put on a magnetic stirrer for 5 h of dissolution at 400 rpm, and the supernatant was taken out after centrifugation. ICP-MS was used to measure the concentrations of Cu, Pb, Zn, and Fe in the solution. The concentrations of Cl^- and SO_4^{2-} were measured by liquid chromatography. For comparison with the component release of pyrite inclusions, seven parallel experiments were performed to obtain the average values. Table 5.23 shows the results of dissolution tests.

TABLE 5.23 Concentrations of Cu, Pb, Zn, Fe, Cl^-, and SO_4^{2-} in the solution from the oxidative dissolution of pyrite.

	Concentration ($\times 10^{-6}$ mol/L)					
Test No.	Cu	Pb[a]	Zn	Fe	Cl^-	SO_4^{2-}
1	0.35	0.69	0.79	2.00	0.98	3.85
2	0.31	1.32	1.24	2.64	2.70	2.28
3	0.15	0.96	1.07	2.97	1.70	2.76
4	0.26	1.56	0.72	4.23	3.10	3.53
5	0.34	1.70	1.12	3.62	2.28	5.60
6	0.28	0.64	0.66	4.31	2.62	3.62
7	0.34	1.28	1.00	3.25	1.80	2.59
Average	0.29	1.16	0.94	2.69	2.17	3.46

[a] $\times 10^{-8}$ mol/L.

Table 5.23 shows the test results of pyrite dissolution, in which the average values of Cu, Pb, Zn, Fe, Cl^-, and SO_4^{2-} in the aqueous solution are 0.29×10^{-6}, 0.94×10^{-8}, 1.16×10^{-8}, 2.69×10^{-6}, 2.17×10^{-6}, and 3.46×10^{-6} mol/L, respectively. The results are compared with the pyrite water wash solution (see Table 5.22). After dissolution of the pyrite, the concentrations of Cu, Fe, Cl^-, and SO_4^{2-} in the aqueous solution were about 10 times lower than those present in the inclusion component release. However, there was no significant difference in the concentrations of Pb and Zn, indicating that Cu, Fe, Cl^-, and SO_4^{2-} from the inclusion fluid are the main sources of unavoidable ions in the solution. The salt solution performs the function of carrying the ore-forming element ions. In theory, the carrying capacity increases with the salinity. These transported ore-forming ions are captured in the mineral crystal defects before they are precipitated during mineralization or, in other cases, they are not precipitated but exist in the form of ions in the enclosed area of fluid inclusions. In addition, the captured fluid is a kind of salt solution, which can act with the interface with the mineral. Some of the ions in the mineral are dissolved into the fluids. These ions will be released during the opening of the inclusions.

5.7.5 Component release of quartz fluid inclusion in the pyrite deposits of Weixin

When different minerals are mineralized at the same period, there are lots of commonalities in the crystal defects and the ore-forming fluids captured. Quartz mineral easily absorbs infrared light due to its good transparency. Thus, the internal structural features can be clearly observed by infrared optical microscopy. Therefore, the study of the inclusions in associated quartz crystals can help us to better understand the inclusions in symbiotic pyrite. The quartz veins in pyrite deposits of Weixin were selected as samples in this study, and IR–UV microscopy was used to analyze the inclusions in the quartz crystals as shown (see Fig. 5.29).

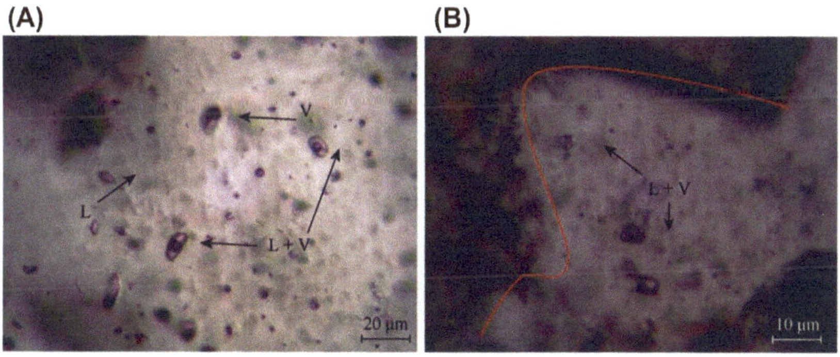

FIGURE 5.29 Infrared optical microscopy images of quartz crystal. *L*, liquid phase; *V*, gas phase; *L + V*, gas–liquid phase; (A) image at low magnification (B) image at high magnification.

Fig. 5.29 shows the infrared imaging results of the quartz crystal surface. There are a large number of fluid inclusions in the quartz crystal, ranging from tens of nanometers to tens of micrometers. The shapes are elongated, elliptical, irregular, and negative crystal; the fluids mainly exist in the form of gas–liquid mixed phase, with a few of pure liquid or pure gas phase. Some small cavity defects are distributed in the microcracks, dislocation nets, and dislocation lines (see the curve in Fig. 5.29B). Some of them are distributed in isolation (See Fig. 5.29A), while some in clusters.

ICP-MS and IC were used to determine the concentrations of the components in the fluid inclusions of the quartz crystals. Table 5.24 shows the concentrations of Cu, Pb, Zn, Fe, Cl^-, and SO_4^{2-} in the aqueous solution after grinding.

Table 5.24 shows that there are large amounts of Cl^- and SO_4^{2-} in the fluid inclusions of quartz, which is consistent with the aforementioned salinity measurement. For the results of the 7 parallel experiments, the average concentrations of Cl^- and SO_4^{2-} in aqueous solution were 46.20×10^{-7} and 27.52×10^{-7} mol/L, respectively. Small amounts of Fe and Cu were detected, and trace amounts of Pb and Zn were also detected with weak individual signals.

The amounts of the components released from the quartz fluid inclusions were compared with the component release from pyrite. The results showed that the released components (namely Cu, Fe, Cl^-, and SO_4^{2-}) of inclusions in the pyrite were also present in the quartz veins of the same region, but in lower concentrations. This may be caused by different amounts of fluid inclusions produced by pyrite and quartz under the same mineralization conditions. The types of components in the inclusions of the quartz were consistent with those of pyrite. That is, the fluids of the inclusions mainly contain Cu, Fe, Cl^-, and

TABLE 5.24 Concentrations of Cu, Pb, Zn, Fe, Cl^-, and SO_4^{2-} in solution released from the macrodefects of quartz.

	Concentration ($\times 10^{-7}$ mol/L)					
Test No.	Cu	Pb[a]	Zn	Fe	Cl^-	SO_4^{2-}
1	0.84	<0.01	<0.01	7.23	23.50	38.80
2	0.26	<0.01	<0.01	10.05	39.45	25.48
3	0.42	<0.01	<0.01	7.90	23.23	20.06
4	0.88	<0.01	<0.01	5.44	35.00	43.54
5	1.01	<0.01	<0.01	10.28	36.57	16.68
6	1.33	<0.01	<0.01	8.29	46.20	27.52
7	0.60	<0.01	<0.01	3.74	66.40	15.35
Average	0.76	<0.01	<0.01	7.56	38.62	26.78

[a] $\times 10^{-8}$ mol/L.

SO_4^{2-}, with no or a very small amount of Pb and Zn. This may be related to the components carried by the ore-forming fluid during diagenesis. Combined with the test results in the previous section, the gangue minerals also release ore-forming element ions in addition to the main target minerals. Thus, the fluid components make an important contribution to the unavoidable ions in the pulp solution, which greatly influences the flotation.

5.7.6 Component release of fluid inclusions of pyrite from the polymetallic sulfide deposits of Dapingzhang

The study of pyrite and quartz samples from Weixin has helped to confirm the existence of fluid inclusions in minerals and also to analyze the characteristics of the inclusions. Meanwhile, quantitative analysis was also conducted to establish the contribution of inclusion components to the unavoidable ions in aqueous solution. The test results of the analysis showed that the fluid inclusions of pyrite and quartz from the Weixin deposits contained few heavy metal elements, mainly Pb and Zn. The content and composition of the inclusions are related to the ore-forming elements and conditions of mineralization. Here, we explored the chemical compositions of the mineral inclusions in polymetallic sulfide deposits. In addition, we explored the relationship between the components in the inclusions and the ore-forming elements. The pyrite and quartz in copper polymetallic deposits from Dapingzhang were selected for this next study.

Fig. 5.30 shows the SEM photograph of the polymetallic ore from Dapingzhang. In the sheets of the associated minerals, there is a paragenetic

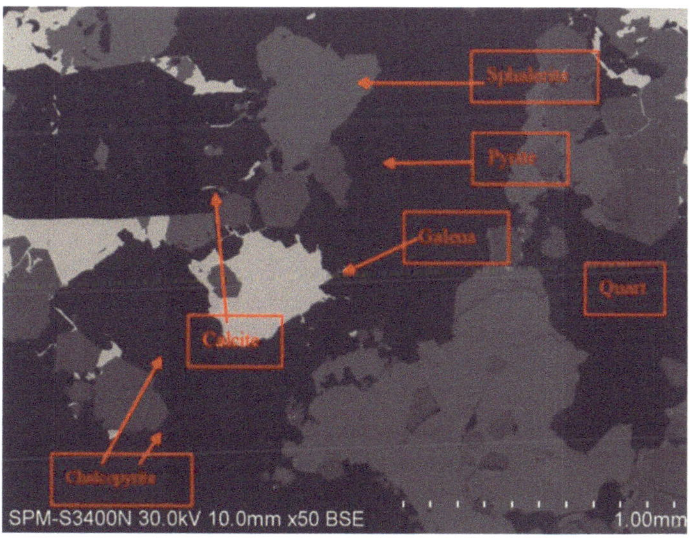

FIGURE 5.30 Paragenetic relationship of the sulfides from the Dapingzhang polymetallic ore.

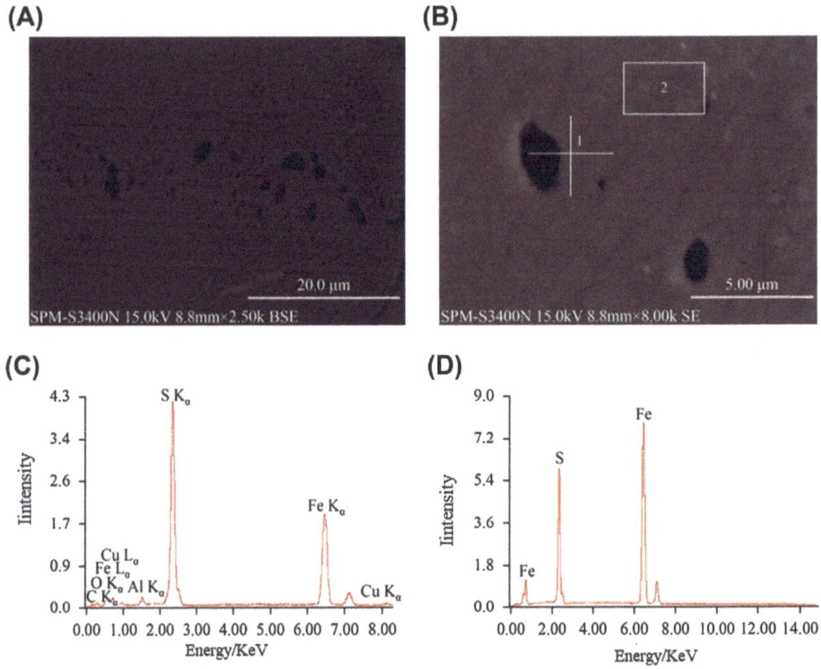

FIGURE 5.31 (A and B) Backscattered electron images of a pyrite slice. (C and D) Energy-dispersive X-ray spectroscopy results. (C) Position 1, (D) position 2. (A) and (B) display the same area at different magnifications.

relationship between the different sulfide minerals (pyrite, chalcopyrite, galena, and sphalerite) and quartz. This paragenetic relationship also exists among the quartz-associated different sulfide minerals, which are leaving the dissolved edges. In addition, they may invade larger cracks in quartz. These ore-forming fluids invade the quartz cracks or dissolve the quartz during the formation of quartz. Observing the contact boundary, some linear fluid inclusions in the quartz cut through the quartz particles and extend to the boundary with the pyrite. Therefore, fluids in such inclusions are related to the conditions of mineralization.

Fig. 5.31 shows the SEM image of the pyrite from Dapingzhang with high magnification of the surface. There are many dimples on the dissociation surface of the pyrite. The structure and morphology reveal there are holes that are traces left after the destruction of the inclusions rather than scratches. These inclusions vary in size and shape. The particle sizes range from a few to tens of micrometers. In addition, the fluid inclusions present a random distribution—some are isolated, while some are concentrated. The shapes include spheres, ellipses, strips, and irregular forms. The fluid inclusions in quartz mainly exist in the form of mixed gaseous–liquid phase, with a small amount of pure liquid or pure gaseous phases.

TABLE 5.25 Concentrations of Cu, Pb, Zn, Fe, Cl^-, and SO_4^{2-} in aqueous solution released from fluid inclusions and dissolution of pyrite.

Project	Test No.	Elements and ions						
		Cu	Pb	Zn	Fe	Ca	Cl^-	SO_4^{2-}
Release of inclusions ($\times 10^{-7}$ mol/L)	1	1.60	16.57	17.93	485.32	23.64	36.00	65.46
	2	3.38	22.57	16.64	506.70	43.52	33.78	32.28
	3	2.31	10.40	23.79	557.54	21.33	35.45	43.44
	Average	2.43	16.51	19.45	516.52	29.50	35.08	47.06
Dissolution ($\times 10^{-9}$ mol/L)	1	0.97	/	7.72	11.10	46.43	5.86	14.52
	2	1.23	/	4.12	12.28	78.28	19.27	10.12
	3	0.50	1.23	1.81	8.62	32.53	17.17	17.88
	Average	0.90	0.41	4.55	10.67	52.41	14.10	14.17

ICP-MS and IC were used to determine the concentrations of Cu, Pb, Zn, Fe, Ca, Cl^-, and SO_4^{2-} in aqueous solutions. They were released and dissolved from pure pyrite inclusions, with the results shown in Table 5.25.

Table 5.25 shows that the concentrations of the ions from dissolution of pyrite are over 2 orders of magnitude lower than those released from the inclusions. This indicates that the released components of fluid inclusions are an important source of the ions in the pulp solution.

5.7.7 Component release of the fluid inclusions of quartz from polymetallic sulfide deposits of Dapingzhang

Fig. 5.32 shows an infrared micrograph of the quartz from Dapingzhang. The image shows that there are many fluid inclusions in quartz, with varied

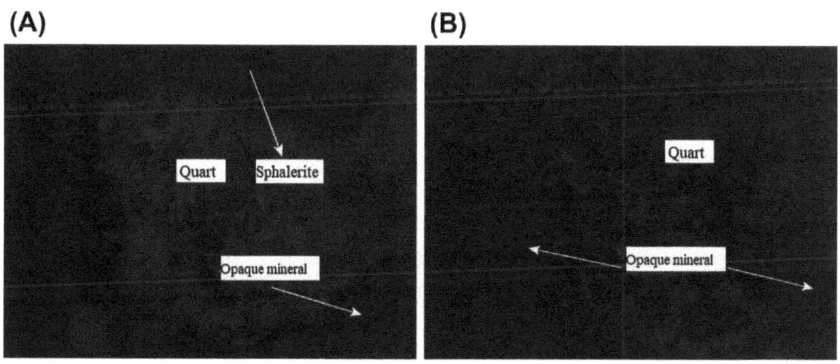

FIGURE 5.32 Infrared optical microscopy images of quartz : (A) image at low magnification (B) image at high magnification.

TABLE 5.26 Concentrations of Cu, Pb, Zn, Fe, Cl^-, and SO_4^{2-} in aqueous solution released from fluid inclusions and nonoxidative dissolution of quartz.

Project	Test No.	Elements and ions						
		Cu	Pb	Zn	Fe	Ca	Cl^-	SO_4^{2-}
Release of inclusions ($\times 10^{-7}$ mol/L)	1	1.95	8.33	7.69	91.86	6.84	23.52	6.32
	2	1.55	9.56	6.23	96.42	4.22	15.23	3.24
	3	2.26	8.75	11.01	82.71	7.58	43.10	2.29
	Average	1.92	8.88	8.31	90.33	6.21	27.28	3.95
Dissolution ($\times 10^{-9}$ mol/L)	1	2.26	/	9.44	0.01	12.52	5.86	/
	2	1.38	/	8.56	/	17.40	19.27	/
	3	1.93	/	7.90	/	/	17.17	/
	Average	1.86	/	8.63	0.01	9.97	14.10	/

sizes and shapes. The particle sizes range from a few to tens of micrometers. In addition, the fluid inclusions present a random distribution—some are isolated, while some are concentrated. The shapes include spheres, ellipses, strips, and irregular forms. Fluid inclusions in this quartz mainly exist in the form of mixed gaseous–liquid phase, with small amounts of pure liquid or pure gas phases. The infrared optical microscopy image shows that other sulfide minerals are also connected to the quartz (see Fig. 5.32A). Some exist in the form of infestation in the sections or cracks of the quartz crystal (see Fig. 5.32B). Fluid inclusions in the quartz intersperse through the linking edges, or through other minerals along the sections and cracks between the minerals. This indicates that the different minerals are formed at the same time, and are capable of capturing fluids to form inclusions. It is likely that they carry the same compositions with different concentrations.

Pure quartz vein samples were adopted to compare the concentrations of the components released from quartz inclusions and those of dissolution of the mineral. We obtained the concentrations of Cu, Pb, Zn, Fe, Ca, Cl^-, and SO_4^{2-} in the aqueous solutions through ICP-MS and IC. Table 5.26 shows that the concentrations of the dissolved ions are over 2 orders of magnitude lower than those released from the inclusions, indicating that the release of fluid inclusions is an important source of the ions in pulp solution. In addition, it also establishes that gangue minerals in polymetallic sulfide deposits contain heavy metal components such as Cu, Pb, Zn, and Fe.

Further reading

[1] Tan KX, Zhang ZR, Wang ZG. The mechanism of surface chemical kinetics of dissolution of minerals. Chin J Geochem 1996;15(1):51–60.

[2] Stanton MR, Gemery-Hill PA, Shanks WC, et al. Rates of zinc and trace metal release from dissolving sphalerite at pH 2.0-4.0. Appl Geochem 2008;23(2):136–47.

[3] Liu J, Wen SM, Xian YJ, et al. Dissolubility and surface properties of a natural sphalerite in aqueous solution. Miner Metall Process 2012;29(2):113–20.

[4] Beaussart A, Mierczynska-Vasilev AM, Harmer SL, et al. The role of mineral surface chemistry in modified dextrin adsorption. J Colloid Interface Sci 2011;357(2):510–20.

[5] Wei Y, Sandenbergh RF. Effects of grinding environment on the flotation of rosh pinah complex Pb/Zn ore. Miner Eng 2007;20(3):264–72.

[6] Rao SR, Finch JA. A review of water re-use in flotation. Miner Eng 1989;2(1):65–85.

[7] Bailly L, et al. Fluid inclusion study of stibnite using infrared microscopy: an example from the brouzils antimony deposit (Vendee, Armorican massif, France). Econ Geol Bull Soc Econ Geol 2000;95(1):221–6.

[8] Luders V, Gutzmer J, Beukes NJ. Fluid inclusion studies in cogenetic hematite, hausmannite, and gangue minerals from high-grade manganese ores in the kalahari manganese field, South Africa. Econ Geol Bull Soc Econ Geol 1999;94(4):589–95.

[9] Campbell AR, et al. Internal features of ore minerals seen with the infrared microscope. Econ Geol 1984;79:1387–92.

[10] Lin C, Miller J. Network analysis of filter cake pore structure by high resolution X-ray microtomography. Chem Eng J 2000;77(1):79–86.

[11] Lin C, Miller J. Cone beam X-ray microtomography-a new facility for three-dimensional analysis of multiphase materials. Miner Metall Process 2002;19(2):65–71.

[12] Lu H, Fan H, Ni P. Fluid inclusions. Beijing: Science Press; 2004.

[13] Wang S, et al. U-Pb age of zirconium ion probe of pulang porphyry copper deposit in western Yunnan: metallogenic time limit and geological significance. Acta Petrol Sin 2008;24(10):2313–21.

[14] Crawford ML. Phase equilibria in aqueous fluid inclusions. In: Fluid inclusions: applications to petrology, vol. 6. Mineralogical Association of Canada; 1981. p. 75–100.

[15] Ding JY, Ni P, Rao B. Synthetic fluid inclusion of $CaCl_2$-H_2O system. Acta Petrol Sin 2005;21(5):1425–8.

[16] Zhai C. The source and ore-forming fluids of dongchuan-yimian copper deposit. J Kunming Univ Sci Technol 1989;4:2.

[17] Hua R. Research on metal precipitation caused by fluid mixing during mineralization. Adv Earth Sci 1994;9(4):15–22.

[18] Bonnemaison M, Marcoux E. Auriferous mineralization in some shear-zones: a three stage model of metallogenesis. Miner Depos 1990;25(2):96–104.

[19] Bortnikov N, et al. The nature of chalcopyrite inclusions in sphalerite; exsolution, coprecipitation, or "disease"? Econ Geol 1991;86(5):1070–82.

[20] Barton PB, Bethke PM. Chalcopyrite disease in sphalerite; pathology and epidemiology. Am Mineral 1987;72(5–6):451–67.

[21] Zhang W. Separation of copper and zinc. Compr Util Miner 1983;2:007.

[22] Qiu X. Theory and application of high-efficiency flotation separation of copper-zinc refractory sulfide ore. Jiangxi University of Science and Technology; 2010.

[23] Peng Y, et al. Control of grinding conditions in the flotation of chalcopyrite and its separation from pyrite. Int J Miner Process 2003;69(1):87–100.

[24] Majima H. How oxidation affects selective flotation of complex sulphide ores. Can Metall Q 1969;8(3):269–73.

[25] Shen WZ, Fornasiero D, Ralston J. Effect of collectors, conditioning pH and gases in the separation of sphalerite from pyrite. Miner Eng 1998;11(2):145–58.

[26] Potter RW, Clynne MA, Brown DL. Freezing point depression of aqueous sodium chloride solutions. Econ Geol 1978;73(2):284–5.

[27] Günther D, Audétat A, Frischknecht R, et al. Quantitative analysis of major, minor and trace elements in fluid inclusions using laser ablation–inductively coupled plasmamass spectrometry. J Anal Atomic Spectrom 1998;13(4):263–70.

[28] Kouzmanov K, Pettke T, Heinrich CA. Direct analysis of ore-precipitating fluids: combined IR microscopy and LA-ICP-MS study of fluid inclusions in opaque ore minerals. Econ Geol 2010;105(2):351–73.

[29] Skou E, Jacobsen T, Van der Hoeven W, et al. On the zinc-chloride complex formation. Electrochim Acta 1977;22(2):169–74.

[30] Wilkinson JJ, Stoffell B, Wilkinson CC, et al. Anomalously metal-rich fluids form hydrothermal ore deposits. Science 2009;323(5915):764–7.

Chapter 6

Solubility of sulfide mineral and chemical behaviors of solution after release of inclusion components

Chapter outline

6.1 Solubility of sulfide mineral **152**
6.1.1 Research methods 152
6.1.2 Solubility of the chalcopyrite surface 153
6.1.3 Solubility of the sphalerite surface 156
6.1.4 Solubility of the pyrite surface 158
6.2 Equilibrium theory calculation of solubility of sulfide minerals **160**
6.2.1 Solubility of sulfide mineral in pure water 160
6.2.2 Solubility of sulfide mineral at different pH 164
6.3 Chemical equilibrium calculation of metal ions in slurry solution **169**
6.3.1 Cu^{2+} taring and component distribution 169
6.3.2 Zn^{2+} taring and component distribution 172
6.3.3 Pb^{2+} taring and component distribution 173
References **177**

According to analyses, there are large numbers of natural fluid inclusions in nonferrous metal (Cu, Pb, and Zn) sulfides and their gangue minerals. The inclusions contain rich chemical components of Na^{+}, K^{+}, Ca^{+}, Mg^{2+}, Cl^{-}, SO_4^{2-}, Cu^{2+}, Pb^{2+}, and Zn^{2+}. Fluid inclusions in minerals are broken and opened up by grinding, leading to the release of "pale fluid" containing the inclusions' components. The large amounts of the released components have an important impact on the chemical system of the pulp and the surface properties of the minerals, especially the activation effects of heavy metal components such as Cu and Pb. Generally, sulfide minerals have low solubility and a small solubility product constant. For example, the solubility of Cu and Zn sulfide minerals ranges from 10^{-9} to 10^{-15} mol/L. The amount of heavy metal components released by the dissolution is limited. Therefore, the heavy metal components released from the inclusions of sulfide minerals and their gangue minerals should be the focus in mineral processing.

Fluid Inclusion Effect in Flotation of Sulfide Minerals. https://doi.org/10.1016/B978-0-12-819845-2.00006-5

In this chapter, the solubility characteristics of sulfide mineral from its self-dissolution are first explained. Furthermore, we analyze the important contributions of the components in fluid inclusions of sulfide mineral to the metal ion system in the pulp. In addition, this chapter focuses on the chemical behavior of the Cu^{2+}, Pb^{2+}, and Zn^{2+} released from mineral inclusions and by oxidative dissolution into the pulp solution. The variations in the solution environment after the components' release can be illustrated by calculation of solutions in chemistry.

6.1 Solubility of sulfide mineral

6.1.1 Research methods

For this book, nonoxidative dissolution (under argon) and oxidative dissolution (under saturated oxygen atmosphere) were performed on the single minerals of Cu, Pb, and Zn sulfide to study their solubility under different conditions. We also determined the amounts of the homonymous metal components produced by the dissolution of the mineral. The components contained are as follows: Cu and Fe produced by dissolution of chalcopyrite, Zn produced by dissolution of sphalerite, Fe produced by dissolution of pyrite, and Pb produced by dissolution of galena.

Deionized water (conductivity of 18 MΩ) used in this experiment was produced by deoxidation for 60 min under high-purity argon (99.99%). The test procedure was as follows: The single mineral was ground to sheets of less than 2 mm. Mineral particles with a size of 1–2 mm were used as samples for the dissolution test. A 0.5% HCl solution was prepared and deoxygenated for mineral cleaning. The samples were soaked in the HCl solution for 12 h. Then they were washed for 10 min, and afterward they were repeatedly rinsed with deionized water 20 times and dried under argon gas protection. Pure minerals were ground to particles sizes of less than 20 μm in a percussive ball mill (MM400, Retsch, from Germany). Two grams of the prepared pure minerals was put into a beaker with 40 mL of deoxygenated deionized water or solution. The beaker was placed on a magnetic stirrer for dissolution test for the required agitation time. After the stirring was completed, the pulp was centrifuged for solid–liquid separation; the supernatant was used for inductively coupled plasma mass spectrometry (ICP-MS). The nonoxidative dissolution test was completed in an argon-protected glove box. During the oxidative dissolution 3 L/min of oxygen was injected into the beaker. The pH values were adjusted with 2 mol/L hydrochloric acid and 2 mol/L sodium hydroxide, to obtain the different pH values of the solution. We used a ZetaProbe potentiometer with acid–base titration function.

Milli-Q 50 ultrapure water and CMOS-grade chemical analysis reagents were adopted to prepare a standard solution to ensure the accuracy and reliability of the ICP-MS test results. The standard solution and the samples were determined by an internal standard method. In other words, 10 mL of a 1×10^{-6} g/L rhodium solution was added to the sample solution and the blank solution as an internal standard, thus reducing or removing the signal interference in the plasma mass spectrometer analysis. The blank sample and sample analysis were performed in the same batch. To reduce operational error of the instruments, the experiment was completed in an ultraclean laboratory.

6.1.2 Solubility of the chalcopyrite surface

When the ground product entered the aqueous solution, hydration occurred on the mineral surface, leading to dissolution of the surface. The ions generated by the dissolution and the surface characteristics after the dissolution have important influences on the flotation. A solubility test was performed to explore the dissolution behavior on the fresh surface of chalcopyrite in aqueous solution. Most of the existing studies have reported the dissolution kinetics of chalcopyrite under conditions of strong oxidation or strong acid and alkali, none of which are the actual flotation medium, water. However, some scholars have studied the long-term solubility of copper minerals in natural water, but the dissolution time was up to hundreds of hours, which did not correspond to the actual flotation time. Therefore, here we studied the dissolution behavior of chalcopyrite in deionized water within a short period of time. Figs. 6.1 and 6.2 show the solution characteristics of chalcopyrite under nonoxidizing and oxidizing environments at different mechanical stirring times and different pH values.

Figs. 6.1 and 6.2 show that at the initial stage, under argon and oxygen atmospheres, the curve of chalcopyrite in aqueous solution has an abnormal point, which is the content of Cu and Fe obtained from the liquid after the freshly ground sample was directly mixed with the aqueous solution. The point occurred repeatedly in the tests. However, the curves showed a certain regular distribution after 6 min of mechanical stirring. In the regular range, the concentrations of Cu and Fe are similar in the aqueous solution, presenting a near-linear relationship with time. In the solutions with strong acidity, basicity, and oxidizing power, different researchers have proposed various chalcopyrite reaction models. There are three stages in the models, including flowing film diffusion, surface chemical reaction, and deep dissolution [3,4]. Each stage has different reaction models, such as $1 - (1 - \alpha)^{1/3} = kt$, $-\ln(1 - \alpha) = kt$, $1 - 2/3\alpha - (1 - \alpha)^{2/3} = kt$, and $[1 - (1 - \alpha)^{1/3}]^2 = kt$, where k is the rate constant.

Many researchers have also reported the effects of mechanical stirring time, particle size, temperature, and oxide concentration in solution on the

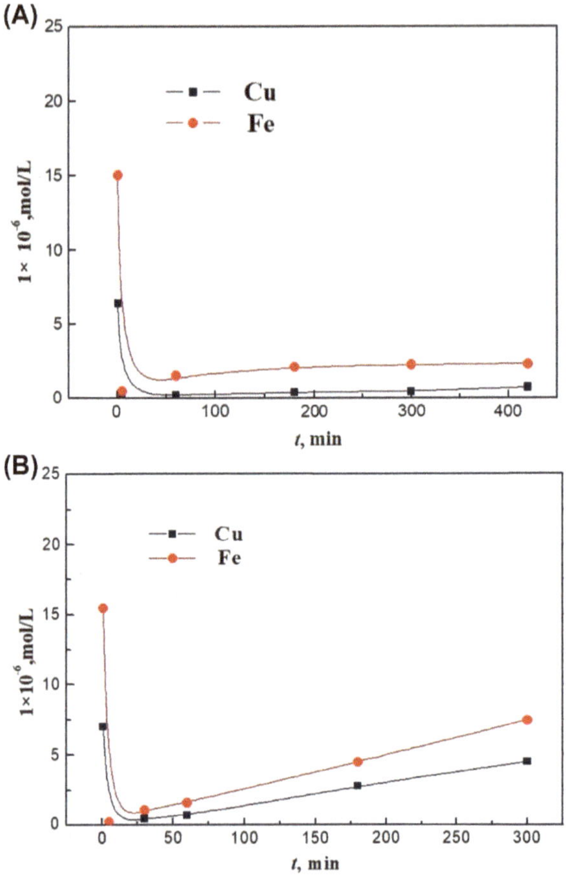

FIGURE 6.1 Plots of the dissolution of chalcopyrite versus time. (A) Under argon atmosphere, (B) under oxygen atmosphere.

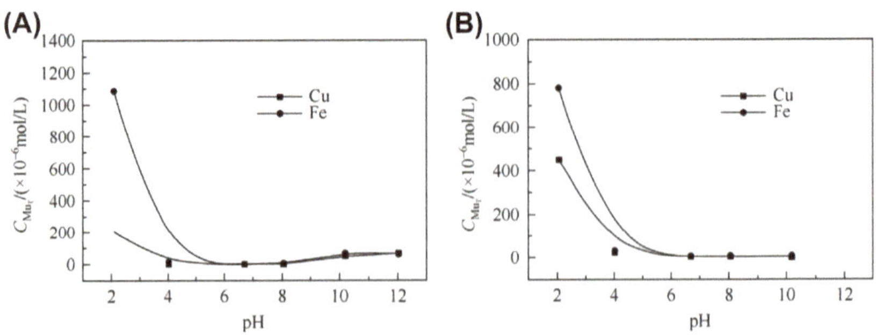

FIGURE 6.2 Plots of the dissolution of chalcopyrite versus pH. (A) Under argon atmosphere, (B) under oxygen atmosphere.

dissolution of chalcopyrite in acidic solution. However, these nonlinear and exponential models are not suitable for the experimental results of this study. The factors of temperature and mechanical stirring time can be kept constant, and the chalcopyrite surface will be changed during the dissolution. The effective specific surface area s has an influence on the dissolution. Taking the dissolution time t as the independent variable, the following model is proposed: $c = \mathrm{k}s^{\mathrm{m}}t + \mathrm{n}$, where k, m, and n are constants. The fitting results show that when $\mathrm{k}s^{\mathrm{m}}$ is constant, the dissolution shows a dynamic equilibrium relationship with time. The natural chalcopyrite surface is dissolved under both argon and oxygen environments. Nevertheless, the results indicate that the effective specific surface area keeps a dynamic balance except for the abnormal points at the initial stage. In the oxygen environment, the linear relationship between the dissolution of Fe on the chalcopyrite surface and the time is relatively poor. This indicates that the surface oxidation has an impact on the dissolution of Fe, thus controlling the dissolution of Fe to some extent. In the oxygen environment, the oxygen acts as an electron acceptor. At 20°C, the overall chemical reaction below is not fierce:

$$CuFeS_{2(s)} + 4O_{2(aq)} = Cu^{2+}_{(aq)} + Fe^{2+}_{(aq)} + 2SO^{2-}_{4(aq)} \tag{6.1}$$

The results show that the early stage of chalcopyrite dissolution is still controlled by the surface properties in the oxygen environment. Surface oxidation reaction does not play a leading role, thus it has little effect on the overall reaction.

In solutions with different pH values under argon and oxygen environments, the solubility of chalcopyrite increases with the acidity of the solution. Under acidic conditions, the overall chemical equation of the system with chalcopyrite is as follows:

$$CuFeS_{2(s)} + 4H^{+}_{(aq)} = Cu^{2+}_{(aq)} + Fe^{2+}_{(aq)} + 2H_2S_{(aq)} \tag{6.2}$$

In neutral and alkaline solutions, chalcopyrite shows similar solubility during the short mechanical agitation times. However, the Fe in the solution is slightly higher than Cu throughout the ranges of pH. The results show that these properties affect the atomic composition and migration to the surface. The dissolution of chalcopyrite can be promoted or hindered by adding acid or base or controlling the pH values of the pulp. Thus, the flotation properties of chalcopyrite and other sulfide minerals can be controlled by adjusting the pH of the solution.

The results show that oxidation has limited effects on the dissolution of chalcopyrite within a short period. It is worth studying the abnormal points in Fig. 6.1, which exist after repeated verification. It will be too late to generate effective dissolution as soon as the ground chalcopyrite enters the solution. Thus, the higher concentrations of Cu and Fe can be attributed only to chalcopyrite. Based on the release of fluid inclusion components during grinding, it is assumed that this point is caused by the fluctuation of the Cu and Fe components in inclusions of chalcopyrite. In addition, the abnormal dissolution of chalcopyrite also explains that the components of the inclusions have been released into the solution. This indicates that the Cu and Fe ions present in the solution resulted mainly from the contents released from the fluid inclusions rather than the dissolution of the chalcopyrite. The concentrations of Cu and Fe in the solution were compared. The results show that under a neutral pH environment the Cu and Fe release from the inclusions in the chalcopyrite makes a major contribution to the contents of Cu and Fe in the solution.

6.1.3 Solubility of the sphalerite surface

To determine the solubility of sphalerite, the dissolution trends of sphalerite were studied under argon and saturated oxygen environments at different pH values and mechanical stirring times. The diagrams in both of the following figures show the results after dissolution, Fig. 6.3 shows the relationship between the total Zn concentration, C_{ZnT}, and pH, and Fig. 6.4 shows the relationship between C_{ZnT} and time. The theoretical calculations in Fig. 6.3 are derived from Eq. (6.3) as follows [5,6]:

$$C_{ZnT}^2 = 2.5 \times 10^{-22} \times \left(1 + 2.51 \times 10^{pH-10} + 2.0 \times 10^{2pH-17} + 1.38 \times 10^{3pH-28} + 3.18 \times 10^{4pH-39}\right) \times \left(1 + 0.70 \times 10^{20-2pH} + 0.77 \times 10^{13-pH}\right) \tag{6.3}$$

The results show that under saturated oxygen and inert atmosphere, the solubility of sphalerite is basically the same at the different pH values and different stirring times. Under acidic conditions, C_{ZnT} increases with the decrease in pH of the solution. Under alkaline conditions, C_{ZnT} rises with the

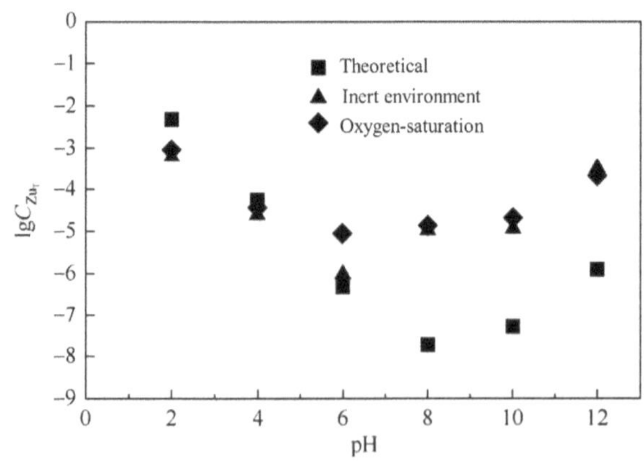

FIGURE 6.3 Relation between $\log(C_{ZnT})$ and pH in solution after dissolution of sphalerite: t = 3 h.

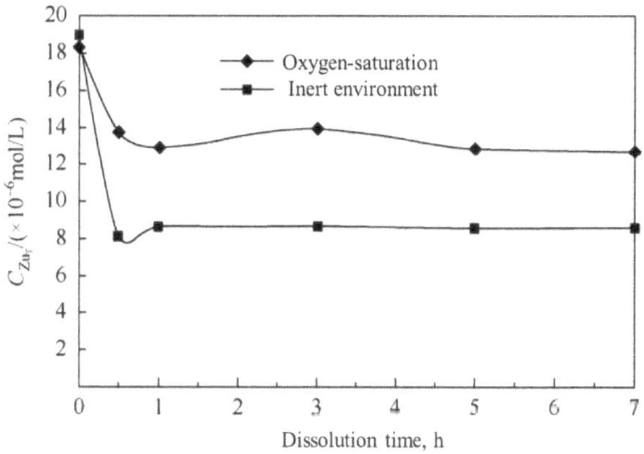

FIGURE 6.4 Relation between log(C_{ZnT}) and pH in solution after dissolution of sphalerite: pH = 6.8.

increase in pH. Thus, more Zn is released into the solution under stronger acidic or alkaline conditions. These relations are consistent with the theoretical calculation trend. Compared with pH, the content of dissolved oxygen in the solution has little effect on the dissolution of sphalerite. Under saturated oxygen and inert conditions, there is no significant change in the concentration of Zn in the solution; the magnitude was consistent. This indicates that sphalerite is stable, not easily dissolved nor oxidized in solution. The dissolution is mainly controlled by the pH of the solution.

It is worth noting that the actual measured C_{ZnT} is consistent with the theoretically calculated order of C_{ZnT} when $pH < 6$, which indicates that the theoretical calculation is reliable. However, when $pH \geq 6$, the concentration of Zn in the sphalerite solution is 100–1000 times larger than the theoretical value in terms of order of magnitude. Such abnormality occurs after repeated measurements. Similar to the dissolution of chalcopyrite, Fig. 6.4 shows that C_{ZnT} measured at the beginning ($t = 0$) of neutral pH is the highest of the whole range of dissolution time. The value is about 1.8×10^{-5} mol/L, which is 1000 times higher than that of the theoretical calculation from Eq. (6.3). It is an abnormal phenomenon. At this time, it is too late for the sphalerite to be dissolved, as it immediately comes in contact with the aqueous solution. It also should be noted that at this point the sphalerite surface does not contain oxidizing components or other impurities from grinding. This higher concentration of Zn at the beginning can be derived only from release of the components of the mineral fluid inclusions.

Fig. 6.3 shows that the amount of Zn dissolved by sphalerite is relatively larger at $pH < 6$, indicating that the release of Zn from fluid inclusions has little effect on C_{ZnT} in the solution. The concentration of Zn dissolved by the sphalerite becomes smaller when $pH \geq 6$. Thus, the Zn released by fluid inclusions contributes to the concentration of Zn in the solution. Similar to the dissolution of chalcopyrite, the oxygen made no significant contribution to the dissolution of the sphalerite within the time range compared with the changes caused by adjusting the pH.

6.1.4 Solubility of the pyrite surface

Figs. 6.5 and 6.6 show the nonoxidative dissolution of pyrite at different pH values, with stirring time of 60 min. The results show that the concentration of S is higher than that of Fe. At pH 2 the result of the nonoxidative dissolution shows that the concentration of S is lower in the solution. The reason is that under strong acidic conditions pyrite reacts with hydrogen ions to release hydrogen sulfide; under weak acidity and weak alkaline conditions the concentration of S in the solution remains almost unchanged. However, in the oxidative dissolution of pyrite the concentration of S increases under strong

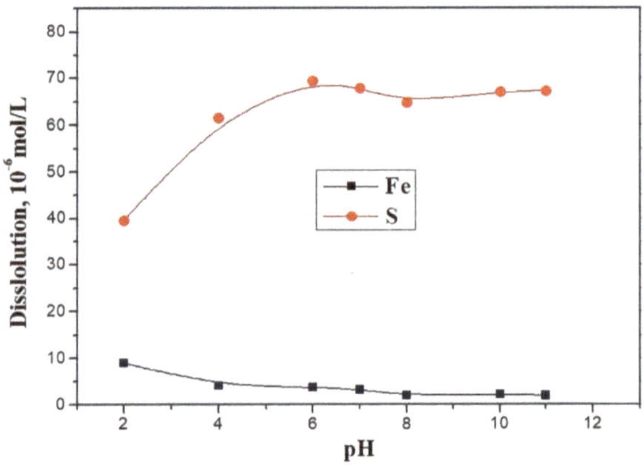

FIGURE 6.5 Plot of the nonoxidative dissolution of pyrite as a function of pH.

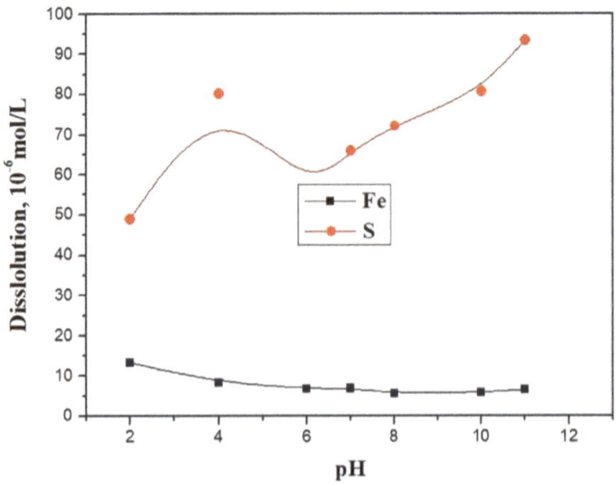

FIGURE 6.6 Plot of the oxidative dissolution of pyrite as a function of pH.

alkaline conditions, because S is oxidized to SO_4^{2-} in the solution. In the nonoxidative and oxidative dissolutions, more Fe was dissolved in the solution under acidic conditions. The concentration of Fe decreases with an increase in the pH value, due to the formation of ferrous hydroxide and iron hydroxide. Thus, the dissolution of pyrite increased in the oxidative dissolution relative to the nonoxidative dissolution [7,8] (see Eqs. (6.4) and (6.5)).

$$FeS_2+7/2O_2+H_2O \rightarrow Fe^{2+}+2SO_4^{2-}+2H^+ \quad (6.4)$$

$$FeS_2+15/4O_2+7/2H_2O \rightarrow Fe(OH)_3+2SO_4^{2-}+2H^+ \quad (6.5)$$

Figs. 6.7 and 6.8 show the relationship between the dissolution and the stirring time of pyrite in a saturated oxygen atmosphere. For both nonoxidative and oxidative dissolution of pyrite, the dissolution amounts of Fe and S were higher at the initial stage (0.2–10 min). Within this time range, the dissolution rate of pyrite decreased with the increase in time, which also occurred in the repeated tests. The dissolution of pyrite presents a certain regularity with further increase in time. In short-term (10–300 min) nonoxidative dissolution of pyrite, the concentration of Fe increases with time; this is accompanied by a nonperiodical fluctuation. In the oxidative dissolution, the concentration of Fe increases with time within 10–120 min, presenting oscillatory variation after 120 min. The pH decreases during the dissolution process, varying in the range of 3.5–4. From the entire dissolution test, the dissolution mechanism of pyrite is very complicated, with lots of intermediate products. In addition, the dissolution of pyrite also shows an abnormal phenomenon in neutral solution. The contents of the Fe and S are higher at the beginning. Then, they decrease with time and finally rise to steady state. The phenomenon of high concentrations of Fe and S at the initial stage and nonperiodical fluctuation during the dissolution process is caused by the release of the fluid inclusion components.

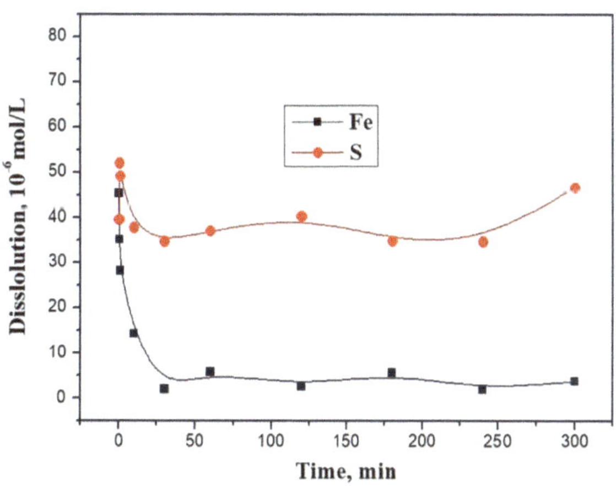

FIGURE 6.7 Plot of the nonoxidative dissolution of pyrite as a function of time.

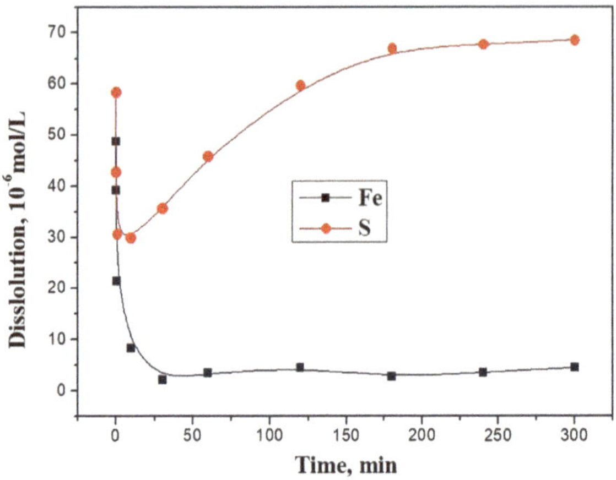

FIGURE 6.8 Plot of the oxidative dissolution of pyrite as a function of time.

6.2 Equilibrium theory calculation of solubility of sulfide minerals

As a common type of mineral in nature, sulfide mineral has a common feature of low solubility in water. However, its solubility is significantly affected by the pH of the solution, which has been demonstrated in the previous study. The oxidative dissolution of sulfide minerals is recognized as an important source of unavoidable ions in ore pulp solutions. In this section, the solubility of sulfide minerals in pure water at different pH values is calculated from the perspective of solution chemistry, thus establishing a complete theoretical system of sulfide mineral dissolution. The solubility of sulfide mineral under different conditions is compared with the concentration of the fluid inclusion contents released from the minerals. We further demonstrate the contribution of the components in mineral fluid inclusions to the pulp during the flotation of sulfide mineral.

6.2.1 Solubility of sulfide mineral in pure water

Sulfide minerals have a certain low level of solubility, unlike most ordinary mineral salts, which have especially high solubility. In saturated aqueous solution, more mineral lattice ions are dissolved, with great impact on the flotation. Some data on the determination of mineral solubility have been generated by most scholars; while they are not wrong, they are not complete either. This is because the solubility of the same mineral may be different in different regions. According to the thermodynamic data, the theoretical values of solubility can be calculated under certain conditions according to the stoichiometry of the minerals and the various equilibrium relationships. On this basis, we discuss the effects of solubility on flotation. In this section, the calculation method was performed following the work of Dianzuo Wang (Wang and Hu, 1988) [8], referring to his book *Solution Chemistry of Flotation*.

For the mineral aqueous solution at equilibrium, it is assumed that all the dissolved components exist in the form of M^{n+} and A^{m-} ions, where M^{n+} is the cation and A^{m-} the anion. Accordingly, the saturated aqueous solution of the mineral at equilibrium contains the following:

$$M_mA_m \rightleftharpoons mM^{n+} + nA^{m-} \tag{6.6}$$

The corresponding equilibrium constant, namely the solubility product of the mineral, is expressed as:

$$K_{sp} = [M^{n+}]^m[A^{m-}]^n \tag{6.7}$$

There are other ligand complexes of M in the solution. The complexation, hydrolysis reaction of M^{n+}, and protonation reaction of A^{m-} have effects on the solubility of the mineral. In this case, the concept of conditional solubility product is used, which is defined as:

$$K'_{sp} = [M]'^m[A]'^n \tag{6.8}$$

where $[M]'$ and $[A]'$ are the total concentrations of M and A in the saturated aqueous solution of the mineral, respectively. $[M]'$ is associated with $[M^{n+}]$ (the free metal ion concentration) by α_m (the side reaction coefficient). Its meaning is roughly the same as that of Eq. 6.8. $[A]'$ is associated with $[A^{m-}]$ by A^{m-} (the side reaction coefficient of the protonation reaction). Therefore:

$$\alpha_M = [M]'/[M^{n+}]\ ,\quad \alpha_A = [A]'/[A^{m-}] \tag{6.9}$$

$$K'_{sp} = [M^{n+}]^m[A^{m-}]^n\alpha_M^m\alpha_A^a = K_{sp}\alpha_M^m\alpha_A^a \tag{6.10}$$

For MA minerals:

$$K'_{sp} = K_{sp}\alpha_M\alpha_A \tag{6.11}$$

The solubility of the mineral is set as S_m, and the unit is mol/L; $[M]' = mS_m$ and $[A]' = nS_m$, then:

$$K'_{sp} = (mS_m)^m(nS_m)^n = K_{sp}\alpha_M^m\alpha_A^a \tag{6.12}$$

$$S_m = \left(\frac{K_{sp}\alpha_M^m\alpha_A^a}{m^m \cdot n^n}\right)^{\frac{1}{m+n}} \tag{6.13}$$

For MA minerals:

$$S_m = (K_{sp}\alpha_M \cdot \alpha_A)^{1/2} \tag{6.14}$$

There are also differences in the hydrolysis and complexation of the cations and the protonation reaction in the different minerals. Thus, we discuss the solubility of the different mineral types as follows.

When the solubility of sulfide minerals is calculated, it is necessary to consider the hydrolysis reaction of the cation and the protonation reaction of the anion S^{2-}. In the saturated aqueous solution of the sulfide mineral, the total concentration of S is:

$$[S'] = [S^{2+}] + [HS^-] + [H_2S]$$

$$= [S^{2+}]\left(1 + K_1^H[H^+] + \beta_2^H[H^+]^2\right)$$

$$\alpha_S = 1 + K_1^H[H^+] + \beta_2^H[H^+]^2 \quad (6.15)$$

The total concentration of metal ions is:

$$[M]' = [M^{n+}] + \left[M(OH)^{n-1}\right] + \cdots \left[M(OH)_k^{n-k}\right]$$

$$\alpha_M = 1 + K_1[OH^-] + \beta_2[OH^-]^2 + \cdots \beta_K[OH^-]^K \quad (6.16)$$

For M_mS_n minerals:

$$K'_{sp} = K_{sp}\alpha_M^m\alpha_S^a \quad (6.17)$$

Taking galena as an example:

$$PbS \rightleftharpoons Pb^{2+} + S^{2-} K_{sp} = 10^{-27.5} \quad (6.18)$$

The change of the solution's pH caused by S^{2-}protonation can be ignored due to the small solubility product of PbS. The pH of the solution is set to 7; then:

$$\alpha_S = 1 + 10^{13.9}[H^+] + 10^{20.92}[H^+]^2 = 1.63 \times 10^7$$

$$\alpha_{1'b} = 1 + 10^{6.3}[OH^-] + 10^{10.9}[OH^-]^2 + 10^{18.9}[OH^-]^3 = 1.2$$

$$K'_{sp} = 10^{-27.5} \times 1.63 \times 10^7 \times 1.2 = 6.19 \times 10^{-21}$$

$$[S_m] = \left[K'_{sp}\right]^{1/2} = 7.87 \times 10^{-11}$$

Similarly, the solubility of other sulfide minerals in pure water can also be calculated according to the relevant data in the following tables (Tables 6.1–6.4) using the above formulas. The values obtained indicate that the solubility of sulfide minerals in pure water is low. The results of the data demonstrate that the calculated solubility is consistent with that reported in most literature.

Although lots of data have been published on the solubility of sulfide minerals in pure water, there are different values, which result from different calculation methods used by the researchers. In most of these studies, the material state has not been the focus, as the different crystal states can cause differences in the solubility of the same mineral. In addition, the types of

TABLE 6.1 Solubility of sulfide minerals in pure water (Wang et al., 1988) [8].

		Solubility (mol/L)				Solubility (mol/L)	
Mineral	Chemical formula	Calculated value	Determined value	Mineral	Chemical formula	Calculated value	Determined value
Covellite	CuS	3.6×10^{-15}		Troilite	FeS	3.6×10^{-8}	
Chalcocite	Cu_2S	1.1×10^{-14}		Greenockite	CdS	1.23×10^{-10}	
Chalcopyrite	$CuFeS_2$	1.9×10^{-14}		Linnaeite	$CoS(\alpha)$	9.0×10^{-8}	
Galena	PbS	7.9×10^{-11}	3.6×10^{-11}	Millerite	$NiS(\alpha)$	8.1×10^{-7}	
Sphalerite	$ZnS(\alpha)$	1.0×10^{-9}	1.47×10^{-9}	Argentite	Ag_2S	1.4×10^{-17}	
Sphalerite	$ZnS(\beta)$			Cinnabar	HgS	5.1×10^{-20}	
Pyrite	FeS_2	5.8×10^{-8}					

TABLE 6.2 Solubility of sulfide minerals and pH value in saturated solution (18–25°C, 100 kPa) [1].

Mineral	Solubility $S/(g\cdot L^{-1})$			pH value in saturated solution
	Literature data	Literature data	Calculated value	
Argentite	8.4×10^{-14}	2.48×10^{-15}	2.5×10^{-12}	7.00
Acanthite	—	—	2.9×10^{-12}	7.00
Chalcocite	1×10^{-13}	1.19×10^{-15}	4.7×10^{-12}	7.00
Covellite	0.00033	2.55×10^{-15}	1.7×10^{-12}	7.00
Troilite	0.0062	0.00616	2.2×10^{-14}	8.22
Pyrite	0.0049	—	1.3×10^{-4}	7.98
Cinnabar	0.00001	0.00001	1.4×10^{-10}	7.00
Metacinnabar	—	—	2.0×10^{-16}	7.00
Alabandite	0.0047	0.00623	3.8×10^{-3}	9.53
Galena	0.00086	9.4×10^{-7}	2.7×10^{-8}	7.00
Wurtzite	0.0069	—	5.3×10^{-6}	7.16
Sphalerite	0.00066	—	4.7×10^{-7}	7.01
Orpiment	0.0005	2.48×10^{-13}	—	—
Bismuthinite	0.00018	0.00018	1.7×10^{-10}	7.00
Bismuthinite	0.00175	0.00175	1.4×10^{-6}	6.98

minerals are different in the literature. Table 6.2 contains some data on the solubility of sulfide minerals in other literature for reference.

6.2.2 Solubility of sulfide mineral at different pH

For the sulfide MS, the equilibrium constant for the dissociation can be expressed as [2]:

$$MS \Leftrightarrow M^{2+} + S^{2-}, \quad K_D = \frac{[M^{2+}][S^{2-}]\gamma_{\pm}^2}{[MS]\gamma_0} \tag{6.19}$$

TABLE 6.3 Stability constant of metal ion hydroxy complex (25°C) Wang et al., 1988) [2].

Metal ion	$\log K_1$	$\log\beta_2$	$\log\beta_3$	$\log\beta_4$	pK_{sp}	$\log\beta_5$
Mg^{2+}	2.58	1.0			11.15	
Ca^{2+}	1.4	2.77			5.22	
Ba^{2+}	0.6				3.6	
Mn^{2+}	3.4	5.8	7.2	7.3	12.6	
Fe^{2+}	4.5	7.4	10.0	9.6	15.1	
Co^{2+}	4.3	8.4	9.7	10.2	14.9	
Ni^{2+}	4.1	8.0	11.0		15.2	
Cu^{2+}	6.3	12.8	14.5	16.4	19.32	
Zn^{2+}	5.0	11.1	13.6	14.8	15.52–16.46	
Pb^{2+}	6.3	10.9	13.9		15.1–15.3	
Cr^{3+}	9.99	11.88		29.87	30.27	
Al^{3+}	9.01	18.7	27.0	33.0	33.5	
Fe^{3+}	11.81	22.3	32.05	34.3	38.8	
Ce^{3+}	5.9	11.7	16.0	18.0	21.9	
Zr^{4+}	14.32	28.26	41.41	55.27	57.2	
La^{3+}	5.5	10.8	12.1	19.1	22.3	
$Ti4^{+}$	14.15	27.88	41.27	54.33	58.3	

If solid MS exists in the solution at saturated state, an improved dissociation constant is obtained as follows:

$$MS(s) \Leftrightarrow MS(aq) \Leftrightarrow M^{2+} + S^{2-} \tag{6.20}$$

where MS(aq) represents the uncharged molecule MS as well as the ion pair $M^{2+}S^{2-}$. In the saturated solution, $[MS]\gamma_0$ is a constant for a particular solute. [MS] includes MS and $M^{2+}S^{2-}$. Eq. (6.19) can be expressed as:

$$[M^{2+}][S^{2-}]\gamma_{\pm}^2 = K_D[MS]\gamma_0 \tag{6.21}$$

TABLE 6.4 Solubility of minerals and compounds (Wang et al., 1988) [8].

Compound	pK_{sp}	Compound	pK_{sp}	Compound	pK_{sp}
MnS (pink)	10.5	$FeCO_3$	10.68	$AlPO_43H_2O$	18.24
MnS (green)	13.5	$ZnCO_3$	10.0	$Ca_{10}(PO_4)_6F_2$	118
FeS	18.1	$PbCO_3$	13.13	$Ca_{10}(PO_4)_6(OH)_2$	115
FeS_2	28.3	$CuCO_3$	9.83	$CaHPO_4$	7.0
CoS(α)	21.3	$CaCO_3$	8.35	$FePO_4 \cdot 2H_2O$	36.0
CoS(β)	25.6	$CaCO_3$	8.22	Fe_2O_3	42.7
NiS(α)	19.4	$MgCO_3$	7.46	FeOH	41.5
NiS(β)	24.9	$CoCO_3$	9.98		16.66
NiS(γ)	26.6				10.41
Cu_2S	48.5	$NiCO_3$	6.87		21.03
CuS	36.1	$CaSO_4$	4.62		18.92
$CuFeS_2$	61.5	$BaSO_4$	9.96		11.08
ZnS(α)	24.7	$PbSO_4$	6.20		13.20
ZnS(β)	22.5	$SrSO_4$	6.50		53.5
CdS	27.0	$CaWO_4$	9.3		50.0
PbS	27.5	$MnWO_4$	8.84		
$MnCO_3$	9.30	$FeWO_4$	11.04		

For saturated solutions, [MS] is the molar solubility or the intrinsic solubility S^0:

$$K_{sp} = [M^{2+}][S^{2-}]\gamma_{\pm}^2 = K_D S^0 \tag{6.22}$$

where γ_0 is defined as 1. If no other equilibrium is involved, the solubility S of this MS electrolyte is the sum of $[M^{2+}]$ or the sum of $[S^{2-}]$ and S_0.

In aqueous solution, M^{2+} is hydrolyzed to form various hydroxyls (hydroxy) or polynuclear hydroxides, causing the complexing effect of hydroxyl (hydrolysis effect):

$$M^{2+} \overset{+OH^-}{\leftrightarrow} M(OH)^+ \overset{+OH^-}{\leftrightarrow} M(OH)_2 \overset{+OH^-}{\leftrightarrow} \cdots \overset{+OH^-}{\leftrightarrow} M(OH)_n^{n-2}$$

S^2 is hydrolyzed to HS^- and H_2S:

$$S^{2-} \overset{+OH^-(pK_{\alpha 2}=12.90)}{\leftrightarrow} HS^- \overset{+H^+(pK_{\alpha 1}=6.97)}{\leftrightarrow} H_2S$$

If ligand L exists in water, L forms a complex with M to increase the solubility. M^{2+} and S^{2-} account for only a small proportion of the total concentration.

Let δ_M^{2+} and δ_S^{2-} be the distribution coefficients of M^{2+} and S^{2-}, respectively, namely the proportions of ions in the total concentration:

$$\delta_M^{2+} = \frac{1}{1 + \sum_{i=1}^{n} \beta_{OH,i}[OH^-]^i + \sum_{i=1}^{m} \beta_{L,i}[L^-]^i} \tag{6.23}$$

$$\delta_S^{2-} = \frac{K_{\alpha 1}K_{\alpha 2}}{[H^+]^2 + [H^+]K_{\alpha 1} + K_{\alpha 1}K_{\alpha 2}} \tag{6.24}$$

where $\beta_{OH,i}$ is the ith cumulative stability constant of M^{2+} and OH^-. $K_{\alpha 1}$ and $K_{\alpha 2}$ are the dissociation constants of H_2S, where $K_{\alpha 1} = 1.3 \times 10^{-7}$ and $K_{\alpha 2} = 7.1 \times 10^{-15}$. $\beta_{OH,i}$ is also the ith cumulative stability constant of M and the ligand L. When the other ligand L is ignored, Eq. (6.23) can be simplified as:

$$\delta_M^{2+} = \frac{1}{1 + \sum_{i=1}^{n} \beta_{OH,i}[OH^-]^i} \tag{6.25}$$

M^{2+} forms polynuclear hydroxide due to hydrolysis. Thus, the calculated values of Eqs. (6.23) and (6.25) are slightly different from the actual measured values. S is set as the solubility of sulfide ($mol \cdot L^{-1}$); $[M^{2+}]$ and $[S^{2-}]$ are the equilibrium concentrations of M^{2+} and S^2; then:

$$[M^{2+}] = S\delta_M^{2+}, \quad [S^{2-}] = S\delta_S^{2-} \tag{6.26}$$

Eq. (6.26) is substituted into Eq. (6.22) to obtain:

$$K_{sp} = S\delta_M^{2+} + S\delta_S^{2-} + \gamma_\pm^2 \tag{6.27}$$

$$S = \sqrt{\frac{K_{sp}}{\gamma_\pm^2 \delta_M^{2+} \delta_S^{2-}}} \tag{6.28}$$

Therefore, for M_2S sulfide:

$$S = \sqrt[3]{\frac{K_{sp}}{4\gamma_\pm^3 \delta_M^{2+} \delta_S^{2-}}} \tag{6.29}$$

For M_2S_3 sulfide:

$$S = \sqrt[5]{\frac{K_{sp}}{108\gamma_\pm^5 \left(\delta_M^{2+}\right)^2 \left(\delta_S^{2-}\right)^3}} \tag{6.30}$$

The simplified Debye–Hückel limit formula is used when the activity coefficient is at 25°C:

$$\log \gamma_\pm = -0.509|Z_+ Z_-|\sqrt{I} \tag{6.31}$$

where $\gamma_\pm$ is the average activity coefficient; 1 is the ionic strength, and $1 = \frac{1}{2}\sum m_i Z_i^2$; the unit is $mol \cdot kg^{-1}$; Z_+ and Z_- are the valences of the cation and anion; m_i is the molar concentration of the ith ion, in the unit of $mol \cdot kg^{-1}$.

The steps to calculate the sulfide solubility at a given pH are as follows:

- **(a)** Substitute pH value into Eqs. (6.24) and (6.25) to calculate δ_M^{2+} and δ_S^{2-}.
- **(b)** Let $\gamma_\pm = 1$.
- **(c)** Substitute K_{sp}, δ_M^{2+}, and δ_S^{2-} into Eq. (6.28), (6.29), or (6.30) to obtain the solubility S_0.
- **(d)** Replace S_0 with m_i to calculate the ionic strength I. SubstituteIinto Eq. (6.31) to calculate $\gamma_\pm$. Then, the solubility S_1 is obtained according to step (c).

In the pH range of 1–12, the solubility S is calculated according to K_{sp} (see Fig. 6.9), showing the solubility rule of sulfide minerals at different pH values. Troilite, alabandite, and wurtzite have higher solubility. For MS-type sulfides, the dissolved S^{2-} is almost completely hydrolyzed to H_2S when $pH \leq 6$. The sulfide gas escapes when the concentration of H_2S reaches $0.1\ mol\ L^{-1}$ (above the dotted line in the figure). The sulfide will continue to dissolve if there is enough acid. Troilite, alabandite, and wurtzite are easily soluble in dilute acid, while argentite, helical silver mineral, and chalcocite are relatively insoluble.

The following conclusions were drawn from the experimental study on the solubility of the aforementioned sulfide minerals and the equilibrium theory

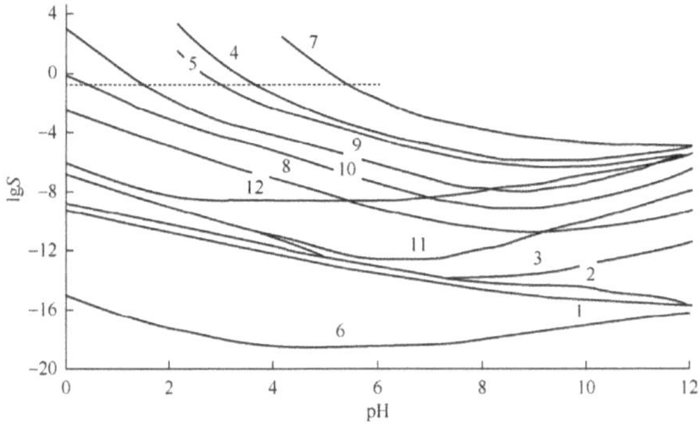

FIGURE 6.9 Relationship between solubility logS and pH. *1*, Argentite and helical silver mineral; *2*, chalcocite; *3*, covellite; *4*, troilite; *5*, pyrite; *6*, cinnabar and metacinnabar; *7*, alabandite; *8*, galena; *9*, wurtzite; *10*, sphalerite; *11*, bismuthinite; *12*, stibnite [1].

calculation. The heavy metal components of Cu, Pb, and Zn released by fluid inclusions play a leading role in the pure aqueous solution. When the pH is changed, the dissolution and release of sulfide mineral makes a major contribution, especially under acidic conditions.

6.3 Chemical equilibrium calculation of metal ions in slurry solution

6.3.1 Cu^{2+} taring and component distribution

The metal ions in minerals undergo a hydrolysis reaction in aqueous solution to form various hydroxyl complexes, where the concentration of each component is determined by the solution equilibrium relationship. Copper ion is the most important unavoidable activated ion in the flotation of nonferrous metal sulfide minerals. The chemical equilibrium and composition distribution of Cu^{2+} in ore pulp solution is the basis of the study of the Cu activation mechanism on sphalerite. The research shows that Cu^{2+} can form four Cu complexes in aqueous solution, namely $Cu(OH)^+$, $Cu(OH)_2$, $Cu(OH)_3^-$, and $Cu(OH)_4^{2-}$. The progressive accumulation constants of the related complexing reactions and other reactions are as follows.

$$Cu^{2+} + OH^- = Cu(OH)^+ \beta_1 = \frac{C_{Cu(OH)^+}}{C_{Cu^{2+}} C_{OH^-}} = 10^{6.3} \tag{6.32}$$

$$Cu^{2+} + 2OH^- = Cu(OH)_2 \beta_2 = \frac{C_{Cu(OH)_2}}{C_{Cu^{2+}} C_{OH^-}^2} = 10^{12.8} \tag{6.33}$$

$$Cu^{2+} + 3OH^- = Cu(OH)_3^- \ \beta_3 = \frac{C_{Cu(OH)_3^-}}{C_{Cu^{2+}}C_{OH^-}^3} = 10^{14.5} \quad (6.34)$$

$$Cu^{2+} + 4OH^- = Cu(OH)_4^{2-} \ \beta_4 = \frac{C_{Cu(OH)_4^{2-}}}{C_{Cu^{2+}}C_{OH^-}^4} = 10^{16.4} \quad (6.35)$$

where β_1, β_2, β_3, and β_4 are the cumulative stability constants of Eqs. (6.32)–(6.35), respectively, and C represents the concentration of the corresponding component. According to mass conservation, the total Cu concentration in the solution can be expressed as:

$$C_{Cu_T} = C_{Cu^{2+}} + C_{Cu(OH)^+} + C_{Cu(OH)_2} + C_{Cu(OH)_3^-} + C_{Cu(OH)_4^{2-}} \quad (6.36)$$

where C_{CuT} represents the total concentration of the various forms of Cu in the aqueous solution. Substituting Eqs. (6.32)–(6.35) into Eq. (6.36), we obtain:

$$C_{Cu_T} = C_{Cu^{2+}}\left(1 + \beta_1 C_{OH^-} + \beta_2 C_{OH^-}^2 + \beta_3 C_{OH^-}^3 + \beta_4 C_{OH^-}^4\right) \quad (6.37)$$

Furthermore, the relationship between the concentration of the hydroxide ions and that of the hydrogen ions can be expressed as:

$$C_{OH^-} = \frac{10^{-14}}{C_{H^+}} \quad (6.38)$$

Substituting Eq. (6.38) into Eq. (6.37), we obtain:

$$C_{Cu_T} = C_{Cu^{2+}}\left(1 + \frac{\beta_1 10^{-14}}{C_{H^+}} + \frac{\beta_2 10^{-28}}{C_{H^+}^2} + \frac{\beta_3 10^{-42}}{C_{H^+}^3} + \frac{\beta_4 10^{-56}}{C_{H^+}^4}\right) \quad (6.39)$$

The side reaction coefficient is defined as α_M, where M = 0–4; α_0, α_1, α_2, α_3, and α_4 represent the percentages of Cu^{2+}, $Cu(OH)^+$, $Cu(OH)_2$, $Cu(OH)_3^-$, and $Cu(OH)_4^{2-}$ in the total C_{CuT} of the solution, respectively. Then, the percentage of each component in the C_{CuT} can be expressed as:

$$\alpha_0 = \frac{C_{Cu^{2+}}}{C_{Cu_T}} = \left(1 + \frac{\beta_1 10^{-14}}{C_{H^+}} + \frac{\beta_2 10^{-28}}{C_{H^+}^2} + \frac{\beta_3 10^{-42}}{C_{H^+}^3} + \frac{\beta_4 10^{-56}}{C_{H^+}^4}\right)^{-1} \quad (6.40)$$

$$\alpha_1 = \frac{C_{Cu(OH)^+}}{C_{CuT}} = \frac{C_{Cu^{2+}}}{C_{CuT}} \times \frac{C_{Cu(OH)^+}}{C_{Cu^{2+}}} = \alpha_0 \frac{\beta_1 10^{-14}}{C_{H^+}} \quad (6.41)$$

$$\alpha_2 = \frac{C_{Cu(OH)_2}}{C_{CuT}} = \frac{C_{Cu^{2+}}}{C_{CuT}} \times \frac{C_{Cu(OH)_2}}{C_{Cu^{2+}}} = \alpha_0 \frac{\beta_2 10^{-28}}{C_{H^+}^2} \quad (6.42)$$

$$\alpha_3 = \frac{C_{Cu(OH)_3^-}}{C_{CuT}} = \frac{C_{Cu^{2+}}}{C_{CuT}} \times \frac{C_{Cu(OH)_3^-}}{C_{Cu^{2+}}} = \alpha_0 \frac{\beta_3 10^{-42}}{C_{H^+}^3} \quad (6.43)$$

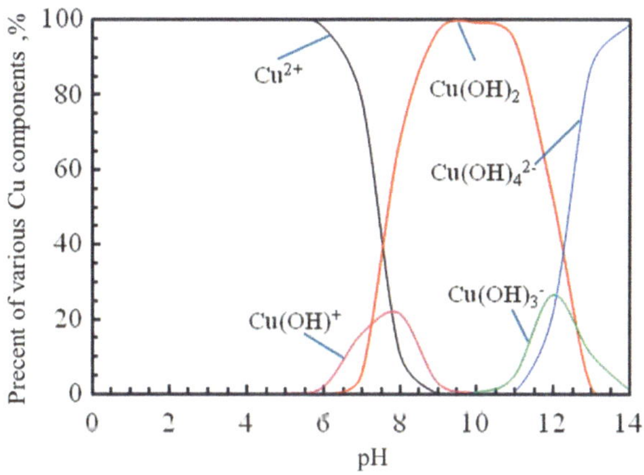

FIGURE 6.10 Relationship between the percentage of the various Cu forms in solution and the solution's pH.

$$\alpha_4 = \frac{C_{Cu(OH)_4^{2-}}}{C_{CuT}} = \frac{C_{Cu^{2+}}}{C_{CuT}} \times \frac{C_{Cu(OH)_4^{2-}}}{C_{Cu^{2+}}} = \alpha_0 \frac{\beta_4 10^{-56}}{C_{H^+}^4} \tag{6.44}$$

The intrinsic relationship among Eqs. (6.40)–(6.44) is:

$$\alpha_0 + \alpha_1 + \alpha_2 + \alpha_3 + \alpha_4 = 1 \tag{6.45}$$

These formulas indicate that the existing form of Cu^{2+} in an aqueous solution is closely related to the pH value of the solution. Substituting the pH value into Eq. (6.40), we obtain the percentage α_0 of free Cu^{2+} in the C_{CuT} at that pH value. Then, the percentages of $Cu(OH)^+$, $Cu(OH)_2$, $Cu(OH)_3^-$, and $Cu(OH)_4^{2-}$ can be obtained by substituting α_0 into Eqs. (6.41)–(6.44). Fig. 6.10 shows the relationship between the relative percentage of the various Cu components and the pH values in the solution.

Fig. 6.10 indicates that Cu^{2+} exists in different forms at the various pH values. When $pH \le 7.5$, Cu exists as free Cu^{2+} and a small amount of $Cu(OH)^+$ and $Cu(OH)_2$, where $Cu(OH)^+$ and $Cu(OH)_2$ appear only when $pH > 5.5$ and 6.5, respectively. Cu exists completely in the form of free Cu^{2+} in the solution when $pH \le 5.5$. When $7.5 < pH \le 12.25$, Cu exists mainly as $Cu(OH)_2$, with small amounts of Cu^{2+}, $Cu(OH)^+$, $Cu(OH)_3^-$, and $Cu(OH)_4^{2-}$. Meanwhile, the content of $Cu(OH)_2$ is over 95% in the range $9 < pH < 11$. When $pH > 12.25$, Cu exists in the form of $Cu(OH)_4^{2-}$ and small amounts of $Cu\ (OH)_3^-$ and $Cu(OH)_2$. The percentage of $Cu(OH)_4^{2-}$ reaches 87% at pH 13.

Fig. 6.10 shows the corresponding relationship between various Cu forms in solution and the pH of the solution. Specific concentration values of Cu^{2+}, $Cu(OH)^+$, $Cu(OH)_2$, $Cu(OH)_3^-$, and $Cu(OH)_4^{2-}$ can be calculated by obtaining the total Cu concentration and pH value of the solution.

6.3.2 Zn^{2+} taring and component distribution

In addition to copper ion, zinc ion is one of the common unavoidable ions in ore pulp solutions of sulfide minerals. The existence of zinc ions is not beneficial to separation flotation of sphalerite, especially for the natural floatability. Hydrophilic Zn hydroxyl complexes are generated when Zn^{2+} combines with OH^- ions in the ore pulp solution. These hydroxyl complexes are adsorbed on the sphalerite surface, thus increasing the hydrophilia of the mineral surface. These reactions are beneficial to flotation of Cu and suppression of Zn during Cu–Zn separation. Studies have shown that the chemical reaction equilibrium of Zn^{2+} is similar to that of Cu^{2+}. In other words, Zn^{2+} also forms four Zn complexes in aqueous solution, including $Zn(OH)^+$, $Zn(OH)_2$, $Zn(OH)_3^-$, and $Zn(OH)_4^{2-}$. Progressive accumulation constants of the relevant coordination reaction and the various reactions are as follows [10,11]:

$$Zn^{2+} + OH^- = Zn(OH)^+ \beta_1 = \frac{C_{Zn(OH)^+}}{C_{Zn^{2+}} C_{OH^-}} = 10^5 \tag{6.46}$$

$$Zn^{2+} + 2OH^- = Zn(OH)_2 \beta_2 = \frac{C_{Zn(OH)_2}}{C_{Zn^{2+}} C_{OH^-}^2} = 10^{11.1} \tag{6.47}$$

$$Zn^{2+} + 3OH^- = Zn(OH)_3^- \beta_3 = \frac{C_{Zn(OH)_3^-}}{C_{Zn^{2+}} C_{OH^-}^3} = 10^{13.6} \tag{6.48}$$

$$Zn^{2+} + 4OH^- = Zn(OH)_4^{2-} \beta_4 = \frac{C_{Zn(OH)_4^{2-}}}{C_{Zn^{2+}} C_{OH^-}^4} = 10^{14.8} \tag{6.49}$$

where $\beta_1, \beta_2, \beta_3$, and β_4 are the cumulative stability constants of Eqs. (6.46)–(6.49), respectively; C represents the concentration of the corresponding components. According to mass conservation, the total Zn concentration in the solution can be expressed as:

$$C_{Zn_T} = C_{Zn^{2+}} + C_{Zn(OH)^+} + C_{Zn(OH)_2} + C_{Zn(OH)_3^-} + C_{Zn(OH)_4^{2-}} \tag{6.50}$$

where C_{ZnT} represents the total concentration of the various forms of Zn in the aqueous solution. Substituting Eqs. (6.46)–(6.49) into Eq. (6.50), we obtain:

$$C_{Zn_T} = C_{Zn^{2+}} \left(1 + \beta_1 C_{OH^-} + \beta_2 C_{OH^-}^2 + \beta_3 C_{OH^-}^3 + \beta_4 C_{OH^-}^4\right) \tag{6.51}$$

Furthermore, the relationship between the concentration of hydroxide ions and that of hydrogen ions can be expressed as:

$$C_{OH^-} = \frac{10^{-14}}{C_{H^+}} \tag{6.52}$$

Substituting Eq. (6.52) into Eq. (6.51), we obtain:

$$C_{Zn_T} = C_{Zn^{2+}}\left(1 + \frac{\beta_1 10^{-14}}{C_{H^+}} + \frac{\beta_2 10^{-28}}{C_{H^+}^2} + \frac{\beta_3 10^{-42}}{C_{H^+}^3} + \frac{\beta_4 10^{-56}}{C_{H^+}^4}\right) \quad (6.53)$$

The side reaction coefficient is defined as α_M, where M = 0–4; α_0, α_1, α_2, α_3, and α_4 represent the percentages of Zn^{2+}, $Zn(OH)^+$, $Zn(OH)_2$, $Zn(OH)_3^-$, and $Zn(OH)_4^{2-}$ in the C_{ZnT} of the solution, respectively. The relevant percentage of each component can be expressed as:

$$\alpha_0 = \frac{C_{Zn^{2+}}}{C_{Zn_T}} = \left(1 + \frac{\beta_1 10^{-14}}{C_{H^+}} + \frac{\beta_2 10^{-28}}{C_{H^+}^2} + \frac{\beta_3 10^{-42}}{C_{H^+}^3} + \frac{\beta_4 10^{-56}}{C_{H^+}^4}\right)^{-1} \quad (6.54)$$

$$\alpha_1 = \frac{C_{Zn(OH)^+}}{C_{ZnT}} = \frac{C_{Zn^{2+}}}{C_{ZnT}} \times \frac{C_{Zn(OH)^+}}{C_{Zn^{2+}}} = \alpha_0 \frac{\beta_1 10^{-14}}{C_{H^+}} \quad (6.55)$$

$$\alpha_2 = \frac{C_{Zn(OH)_2}}{C_{ZnT}} = \frac{C_{Zn^{2+}}}{C_{ZnT}} \times \frac{C_{Zn(OH)_2}}{C_{Zn^{2+}}} = \alpha_0 \frac{\beta_2 10^{-28}}{C_{H^+}^2} \quad (6.56)$$

$$\alpha_3 = \frac{C_{Zn(OH)_3^-}}{C_{ZnT}} = \frac{C_{Zn^{2+}}}{C_{ZnT}} \times \frac{C_{Zn(OH)_3^-}}{C_{Zn^{2+}}} = \alpha_0 \frac{\beta_3 10^{-42}}{C_{H^+}^3} \quad (6.57)$$

$$\alpha_4 = \frac{C_{Zn(OH)_4^{2-}}}{C_{ZnT}} = \frac{C_{Zn^{2+}}}{C_{ZnT}} \times \frac{C_{Zn(OH)_4^{2-}}}{C_{Zn^{2+}}} = \alpha_0 \frac{\beta_4 10^{-56}}{C_{H^+}^4} \quad (6.58)$$

Similarly, after substituting the pH value into Eq. (6.54), we obtain the percentage α_0 of Zn^{2+}. Then, the percentages of $Zn(OH)^+$, $Zn(OH)_2$, $Zn(OH)_3^-$, and $Zn(OH)_4^{2-}$ can be obtained by substituting α_0 into Eqs. (6.55)–(6.57). Fig. 6.11 shows the relationship between the relative percentage of the various Zn components and the pH values in the solution.

Fig. 6.11 shows that when pH ≤ 8.5, Zn exists mainly as free Zn^{2+} with small amounts of $Zn(OH)^+$ and $Zn(OH)_2$. $Zn(OH)^+$ and $Zn(OH)_2$ appear at pH > 6.5. Zn exists in the form of free Zn^{2+} in the solution when pH < 6.5. The percentage (about 9%) of $Zn(OH)^+$ is the largest when pH is 8–8.5. Zn exists mainly as $Zn(OH)_2$ when 8.5 < pH < 11.5, with small amounts of Zn^{2+}, $Zn(OH)^+$, and $Zn(OH)_3^-$. When 11.5 < pH < 13, Zn exists mainly in the form of $Zn(OH)_3^-$, reaching a maximum of 68% at pH 12.25. $Zn(OH)_4^2$ is the form of Zn in the solution at pH >13.

6.3.3 Pb^{2+} taring and component distribution

In aqueous solution, the heavy metal ion Pb^{2+} will form complex ions of $Pb(OH)^+$, $Pb(OH)_2$, $Pb(OH)_3^-$, $Pb_2(OH)^{3+}$, $Pb_4(OH)_4^{4+}$, and $Pb_6(OH)_8^{4+}$.

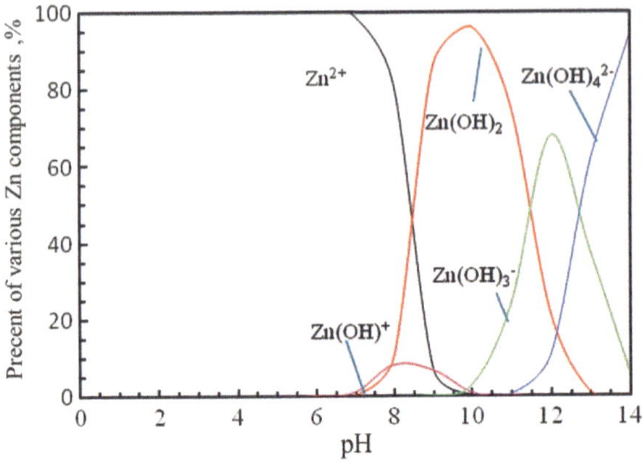

FIGURE 6.11 Percentage of various Zn components versus pH in solution.

The related complexing reactions and progressive accumulation constants are as follows [12]:

$$Pb^{2+} + OH^- \rightleftharpoons Pb(OH)^+ \beta_1 = \frac{[Pb(OH)^+]}{[Pb^{2+}][OH^-]} = 10^{6.2} \tag{6.59}$$

$$Pb^{2+} + 2OH^- \rightleftharpoons Pb(OH)_2\ \beta_2 = \frac{[Pb(OH)_2]}{[Pb^{2+}][OH^-]^2} = 10^{10.3} \tag{6.60}$$

$$Pb^{2+} + 3OH^- \rightleftharpoons Pb(OH)_3^-\ \beta_3 = \frac{[PbOH_3^-]}{[Pb^{2+}][OH^-]^3} = 10^{13.3} \tag{6.61}$$

$$2Pb^{2+} + OH^- \rightleftharpoons Pb_2(OH)^{3+}\ \beta_4 = \frac{[Pb_2(OH)^{3+}]}{[Pb^{2+}]^2[OH^-]} = 10^{7.6} \tag{6.62}$$

$$4Pb^{2+} + 4OH^- \rightleftharpoons Pb_4(OH)_4^{4+}\ \beta_5 = \frac{[Pb_4(OH)_4^{4+}]}{[Pb^{2+}]^4[OH^-]^4} = 10^{36.1} \tag{6.63}$$

$$6Pb^{2+} + 8OH^- \rightleftharpoons Pb_6(OH)_8^{4+}\ \beta_6 = \frac{[Pb_6(OH)_8^{4+}]}{[Pb^{2+}]^6[OH^-]^8} = 10^{69.3} \tag{6.64}$$

From Eq. (6.59), it can also be expressed as:

$$\frac{[Pb(OH)^+]}{[Pb^{2+}]} = \beta_1[OH^-] \tag{6.65}$$

Similarly, Eqs. (6.66)–(6.70) can be expressed from Eqs. (6.60)–(6.64), respectively, as:

$$\frac{[Pb(OH)_2]}{[Pb^{2+}]} = \beta_2[OH^-]^2 \tag{6.66}$$

$$\frac{\left[PbOH_3^-\right]}{[Pb^{2+}]} = \beta_3[OH^-]^3 \tag{6.67}$$

$$\frac{\left[Pb_2(OH)^{3+}\right]}{[Pb^{2+}]^2} = \beta_4[OH^-] \tag{6.68}$$

$$\frac{\left[Pb_4(OH)_4^{4+}\right]}{[Pb^{2+}]^4} = \beta_5[OH^-]^4 \tag{6.69}$$

$$\frac{\left[Pb_6(OH)_8^{4+}\right]}{[Pb^{2+}]^6} = \beta_6[OH^-]^8 \tag{6.70}$$

Therefore, the total lead content in aqueous solution can be expressed as:

$$\begin{aligned}[Pb]_T = {} & [Pb^{2+}] + [Pb(OH)^+] + [Pb(OH)_2] + \left[PbOH_3^-\right] \\ & + 2\left[Pb_2(OH)^{3+}\right] + 4\left[Pb_4(OH)_4^{4+}\right] + 6\left[Pb_6(OH)_8^{4+}\right]\end{aligned} \tag{6.71}$$

The side reaction coefficient is defined as:

$$\alpha_0 = \frac{[Pb^{2+}]}{[Pb]_T} \tag{6.72}$$

$$\alpha_1 = \frac{[Pb(OH)^+]}{[Pb]_T} \tag{6.73}$$

$$\alpha_2 = \frac{[Pb(OH)_2]}{[Pb]_T} \tag{6.74}$$

$$\alpha_3 = \frac{\left[PbOH_3^-\right]}{[Pb]_T} \tag{6.75}$$

$$\alpha_4 = \frac{2\left[Pb_2(OH)^{3+}\right]}{[Pb]_T} \tag{6.76}$$

$$\alpha_5 = \frac{4\left[Pb_4(OH)_4^{4+}\right]}{[Pb]_T} \tag{6.77}$$

$$\alpha_6 = \frac{6\left[Pb_6(OH)_8^{4+}\right]}{[Pb]_T} \tag{6.78}$$

Substituting Eqs. (6.65), (6.70), and (6.72) into Eqs. (6.73) and (6.78), respectively, we obtain:

$$\alpha_1 = \frac{[Pb(OH)^+]}{[Pb]_T} = \frac{[Pb^{2+}]}{[Pb]_T} \times \frac{[Pb(OH)^+]}{[Pb^{2+}]} = \alpha_0 \cdot \beta_1[OH^-] \tag{6.79}$$

$$\alpha_2 = \frac{[Pb(OH)_2]}{[Pb]_T} = \frac{[Pb^{2+}]}{[Pb]_T} \times \frac{[Pb(OH)_2]}{[Pb^{2+}]} = \alpha_0 \cdot \beta_2[OH^-]^2 \tag{6.80}$$

$$\alpha_3 = \frac{\left[PbOH_3^-\right]}{[Pb]_T} = \frac{[Pb^{2+}]}{[Pb]_T} \times \frac{\left[PbOH_3^-\right]}{[Pb^{2+}]} = \alpha_0 \cdot \beta_3[OH^-]^3 \tag{6.81}$$

$$\alpha_4 = \frac{2\left[Pb_2(OH)^{3+}\right]}{[Pb]_T} = \frac{[Pb^{2+}]^2}{[Pb]_T} \times \frac{2\left[Pb_2(OH)^{3+}\right]}{[Pb^{2+}]^2} = 2\alpha_0^2 \cdot \beta_4[OH^-] \cdot [Pb]_T \tag{6.82}$$

$$\alpha_5 = \frac{4\left[Pb_4(OH)_4^{4+}\right]}{[Pb]_T} = \frac{[Pb^{2+}]^4}{[Pb]_T} \times \frac{4\left[Pb_4(OH)_4^{4+}\right]}{[Pb^{2+}]^4} = 4\alpha_0^4 \cdot \beta_5[OH^-]^4 \cdot [Pb]_T^3 \tag{6.83}$$

$$\alpha_6 = \frac{6\left[Pb_6(OH)_8^{4+}\right]}{[Pb]_T} = \frac{[Pb^{2+}]^6}{[Pb]_T} \times \frac{6\left[Pb_6(OH)_8^{4+}\right]}{[Pb^{2+}]^6} = 6\alpha_0^6 \cdot \beta_6[OH^-]^8 \cdot [Pb]_T^5 \tag{6.84}$$

By substituting Eqs. (6.79) and (6.84) into Eq. (6.71), we obtain:

$$1 = \alpha_0 + \alpha_0 \cdot \beta_1[OH^-] + \alpha_0 \cdot \beta_2[OH^-]^2 + \alpha_0 \cdot \beta_3[OH^-]^3 + 2\alpha_0^2 \cdot \beta_4[OH^-] \cdot [Pb]_T + 4\alpha_0^4 \cdot \beta_5[OH^-]^4 \cdot [Pb]_T^3 + 6\alpha_0^6 \cdot \beta_6[OH^-]^8 \cdot [Pb]_T^5 \tag{6.85}$$

These formulas indicate that the concentration fraction of various ions is closely related to pH and total lead ion concentration. Fig. 6.12 shows the relationship between the relative percentage of the various Pb components and the pH values in the solution.

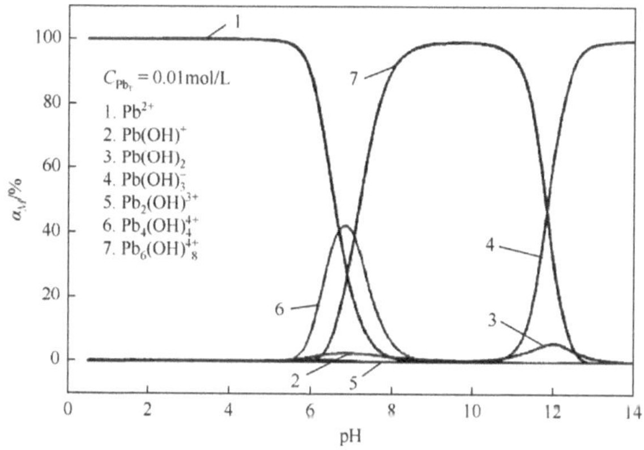

FIGURE 6.12 Relationship between percentage of various Pb components and pH in solution [12].

References

[1] Jia J, Lan B, Xie X, et al. Relationship between solubility of sulfide minerals and pH value of solution. J Changchun Univ Sci Technol 2001;31(3):241–6.

[2] Hu Y. Flotation solution chemistry. Met Ore Dress Abroad 1990;27(11):1–7.

[3] Ikiz D, Gülfen M, Aydın A. Dissolution kinetics of primary chalcopyrite ore in hypochlorite solution. Miner Eng 2006;19(9):972–4.

[4] Prosser AP. Review of uncertainty in the collection and interpretation of leaching data. Hydrometallurgy 1996;41(2):119–53.

[5] Liu J, Wen S, Xian Y, et al. Dissolubility and surface properties of a natural sphalerite in aqueous solution. Miner Metall Process 2012;29(2).

[6] Lu L. The study oil surface reaction of pyrite. Ph D. Thesis. Nanjing University; 2002 (in Chinese with English abstract).

[7] Singer PC, Stumm W. Acid mine drainage: the rate limiting step. Science 1970;167:1121–3.

[8] Moses CO, Nordstrom DK, Herman JS, Mills AL. Aqueous pyrite oxidation by dissolved oxygen and ferric iron. Geochem Cosmochim Acta 1987;51:1561–71.

[9] Wang D, Hu Y. Solution Chemistry of Flotation. Hunan science and Technology Press; 1988.

[10] Veeken AHM, Akoto L, Hulshoff Pol LW, et al. Control of the sulfide (S2-) concentration for optimal zinc removal by sulfide precipitation in a continuously stirred tank reactor. Water Res 2003;37(15):3709–17.

[11] Li E, Pei X, Shu Y, et al. Thermodynamic equilibrium of Zn^{2+}-S2–H_2O system. Environ Sci Technol 2010;33(3):1–3.

[12] Feng N. Chemical modification of orange peel and its adsorption behavior on heavy metal ions. Central South University; 2009.

Chapter 7

Interactions among components of fluid inclusions in sulfide mineral, mineral surfaces, and collectors

Chapter outline

7.1 ζ potential measurement of the adsorption of released components of inclusions on mineral surfaces 180

7.1.1 Test method for ζ potential 181

7.1.2 Surface adsorption of inclusion components in chalcopyrite 182

7.1.3 Surface adsorption of inclusion components in sphalerite 186

7.1.4 Surface adsorption of inclusion components in pyrite and galena 188

7.2 Density functional theory of the interactions between the components of fluid inclusion colonies and mineral surfaces 190

7.2.1 Crystal model and calculation method 190

7.2.1.1 Chalcopyrite crystal model and calculation method 191

7.2.1.2 Sphalerite crystal model and calculation method 193

7.2.1.3 Pyrite crystal model and calculation method 194

7.2.2 Surface relaxation and reconstitution of sulfide mineral 195

7.2.2.1 Surface structure of chalcopyrite 196

7.2.2.2 Surface structure of sphalerite 201

7.2.2.3 Surface structure of pyrite 206

7.2.3 Interaction of chalcopyrite surface and the copper in inclusions 209

7.2.4 Interaction of sphalerite surface and the copper in inclusions 211

7.2.5 Interaction of the pyrite surface and the Cu in inclusions 218

7.3 Interactions among sulfide mineral surface, components of fluid inclusions, and collectors 219

Fluid Inclusion Effect in Flotation of Sulfide Minerals. https://doi.org/10.1016/B978-0-12-819845-2.00007-7

7.3.1 Interaction of chalcopyrite surface, Cu components, and xanthate 220
7.3.2 Interaction of sphalerite surface, Cu, and xanthate 223
7.4 Interaction of the components released from mineral fluid inclusions in sulfide mineral flotation 226
Reference 229
Further reading 229

According to the introduction to this book, the release of the components of fluid inclusions in nonferrous metal sulfide minerals and their gangue minerals is an important new source of inevitable metal ions in ore pulp solutions. Especially in the natural pH of the ore pulp system, the component release of the inclusions predominates. For mineral flotation, fluid inclusion components have two major effects. The first is the influence of the inclusion components on the ore pulp solution system, which has been discussed in Chapter 6. The other is the effect of the inclusion components on the surface properties of the minerals and collectors, which is achieved by the interaction of the mineral surface and certain components of the inclusions with specific surface affinities (such as Cu and Pb, which have an activating effect).

The interactions between inclusion components and mineral surfaces are studied from both macroscopic and microscopic perspectives. From the macroscopic study, we learn that when the metal ions in the inclusions interact with the mineral surface, the electrical properties of the mineral surface are changed. Therefore, this can be characterized by the change rule of the zeta (ζ) potential of the mineral surface in the ore pulp system after grinding. From the microcosmic aspect, density functional theory (DFT) in quantum chemical calculations is used, with Cu (typical heavy metal activation component) as a representative of the inclusion components. This chapter introduces the microscopic mechanism of the interaction between the Cu in the inclusion components and the sulfide mineral surface, as well as the ternary system among the sulfide mineral surface, Cu, and the collector. In addition, we clarify the effects of the release of fluid inclusion components on sulfide mineral flotation.

7.1 ζ potential measurement of the adsorption of released components of inclusions on mineral surfaces

In ore pulp flotation, the mineral surface is charged by the surface group, the dissociation of ions, or the selective adsorption of certain ions from solution. According to the rule of electrical neutrality, there must be ions (which are called counterions) of equal and opposite magnitude of charge at the liquid–solid interphase (solution and mineral particle). The mineral surface forms a double layer with the counterions. When the mineral particles and the liquid-phase diffusion layer move relative to each other, a sliding surface is generated. The potential difference at that point in the fluid away from the interface on the sliding surface is called the ζ potential or electrokinetic potential. Therefore, the ζ potential reflects the surface electrical properties of the charged particles in the solution system, affected by pH, mineral surface properties, type of ion exchange, concentration of the ions, solution dielectric constant, and temperature. Studying the relationship between the ζ potential change of mineral particles and the adsorption properties in solution can be used to explore the adsorption process of the mineral surface and its ion exchange law.

As discussed in Chapter 6, there are reversible processes in the ore pulp solution according to solution chemistry, for example, the hydrolysis reactions and physicochemical adsorption on the mineral surfaces. Under constant conditions, there is a stable process in these reactions; that is, with the equal rates of positive and negative reactions, the reaction in the solution and the ion concentration reach an equilibrium state. The type, state, and concentration of the ions in the solution are important factors affecting the adsorption of the minerals, which in turn change the ζ potential. Thus, when there are adsorbed ions in the ore pulp solution, the ζ potential of the mineral particles changes with the ion species and concentrations in the solution, and eventually tends to be in equilibrium. Therefore, during grinding, the released components of mineral inclusions are adsorbed on the mineral surface. This adsorption can be characterized and explained by the change rule of the ζ potential of the mineral surface after release of the mineral inclusions.

7.1.1 Test method for ζ potential

The ζ potential was measured by the ZetaProbe interface potential analyzer of Colloidal Dynamics (USA), which is one of the most accurate interface potential analyzers. The conductivity, potential, pH, and temperature were calibrated with standard solution before the use of the instrument, to ensure the accuracy of measurement. After calibration, the ζ potential is measured by inputting the density, dielectric constant, mass percentage of the mineral sample, viscosity, and dielectric constant of pure water in the Instrument Operation tab. The procedure for the test was as follows: minerals with particle size of about 1 mm were washed five times with deionized water and then dried in a glove box under argon atmosphere. Fourteen grams of dried pure mineral was taken in each test and placed in two same-model mill pots. The closed mill pots were then mounted on a percussive ball mill (MM400, Retsch, Germany), with impacting frequency of 900 min^{-1}, and ground for 8 min. Each test took 14 g of the mineral samples and milled it using the ball mill. The sample from the meal was placed quickly in a measuring cup, and then 275 mL of ionized water was added. The stirring button of the ZetaProbe was turned on, and the rotary speed was set to 200 rpm. Potentiometric titration was carried out using 2 mol/L hydrochloric acid and 2 mol/L sodium hydroxide solutions as the titration solutions.

7.1.2 Surface adsorption of inclusion components in chalcopyrite

The isoelectric point (IEP) of a pure mineral solution of chalcopyrite was determined, showing that the IEP of the chalcopyrite used in the test was pH 2.4, which is consistent with the reported IEP value of chalcopyrite, between pH 2 and 3. Fig. 7.1 shows the ζ potential results, under ideal working conditions and test materials. The fresh chalcopyrite samples after grinding were directly placed in a ζ potential measuring cup.

Ions undergo various reactions in aqueous solutions, such as hydrolysis reactions, adsorption on mineral surfaces, and chemical reactions. These reactions proceed in the most favorable direction until the concentration of the opposing ions in the solution reaches equilibrium. Factors such as ion species, state, and concentration in the solution affect the ζ potential of the mineral. That is to say, in aqueous solution, the ζ potential of the mineral changes with the ion species and the concentration of the species in solution, which also attain equilibrium after some time. The ionic components of the solution are affected by the components in the aqueous solution, the oxidative dissolution of the mineral surface, and the component release of fluid inclusions in minerals. The fluid inclusions in chalcopyrite contain a large amount of electrolyte solution remaining in mineralization, including homonymous copper and iron ions, as well as chloride salt and sulfate.

From the previous chapter (chapter 5), the concentrations of Cu and Fe released from the fluid inclusions of chalcopyrite were 5.79×10^{-6} and 17.20×10^{-6} mol/L, respectively. These ions (Cu and Fe) were adsorbed on the mineral surface, thus affecting the electrical properties of the chalcopyrite surface. In Fig. 7.1, the ζ potential shows a turbulence with the complex reaction in the solution, but the overall trend of the ζ potential can be observed. In the first 0–5 min, the potential value increased from −1.13 to −0.5 mV.

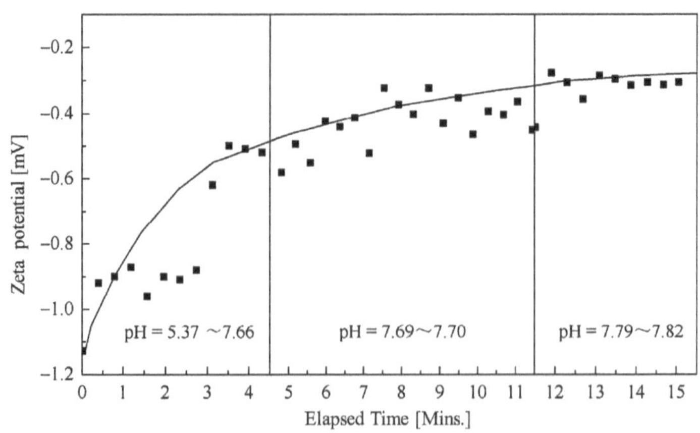

FIGURE 7.1 Change in ζ potential with time in an aqueous solution of chalcopyrite.

The pH starting value was 5.37, and the acidity of the solution was mainly caused by the hydrolysis of metal ions. Taking copper ions as an example, the main reactions in the solution are as follows:

$$H_2O_{(aq)} \leftrightarrows OH_{(aq)}{}^- + H_{(aq)}{}^+ \tag{7.1}$$

$$Cu_{(aq)}{}^{2+} + OH_{(aq)}{}^- \leftrightarrows Cu(OH)_{(aq)}{}^+ \tag{7.2}$$

$$Cu(OH)_{(aq)}{}^+ + OH_{(aq)}{}^- \leftrightarrows Cu(OH)_{2(aq)} \tag{7.3}$$

$$Cu(OH)_{2(aq)} \leftrightarrows Cu(OH)_{2(s)} \tag{7.4}$$

However, the pH in the test rose, which meant that the solution was not acidic. The concentration of the metal ions in the aqueous solution was lowered as the hydrolysis decreased. Thus, the decrease in metal ions should be attributed to the adsorption of ions on the mineral surfaces. This process affected the relative motion of the mineral particles and solution, which had an influence on the ζ potential. At 7–20 min, the potential value rose slowly, and reached equilibrium at 18 min, with a potential of −0.3 mV. The pH 7.7 was maintained from 7 to 20 min with little change, indicating that the various reactions in the solution reached equilibrium. The overall trend of the ζ potential gradually rose until a stable process was attained, indicating that inevitable ions existed in the solution, and there was adsorption of the ions on the mineral surfaces. The chalcopyrite dissolved during the mineral grinding, and chalcopyrite surfaces are rich in S due to relaxation and structural reconstruction. It was speculated that components of the chalcopyrite inclusions would interact with the surface S via the released metal ions.

To prove that copper ions in the released components of the fluid inclusions of chalcopyrite could be adsorbed on the chalcopyrite surface, the experiment was conducted with artificial addition of copper ions. One gram of fresh chalcopyrite sample was weighed out and ground by ball mill. Then 20 mL of 1.10×10^{-4} mol/L Cu^{2+} solution was placed in a closed vial for adsorption tests at different times. Solid–liquid separation was carried out by centrifuge, and inductively coupled plasma mass spectrometry was used to test for the residual Cu element in the supernatant, while X-ray photoelectron spectroscopy (XPS) analysis was carried out on the solid samples for 28 min. Figs. 7.2 and 7.3 and Table 7.1 show the test results.

The copper concentration (C_{Cu}) in the solution decreased from 110×10^{-6} to 46.4×10^{-6} mol/L in 0.5 min. It decreased to 2.7×10^{-6} mol/L by 3 min, and to 1.3×10^{-6} mol/L after 6 min. The comparison shows the largest decline in the first 3 min. After adsorption for 15 min, the C_{Cu} in the solution was basically unchanged, indicating that the reaction of copper ions in the aqueous solution reached equilibrium. The decrease in copper ions in the solution indicates the adsorption of copper ions on the chalcopyrite surface.

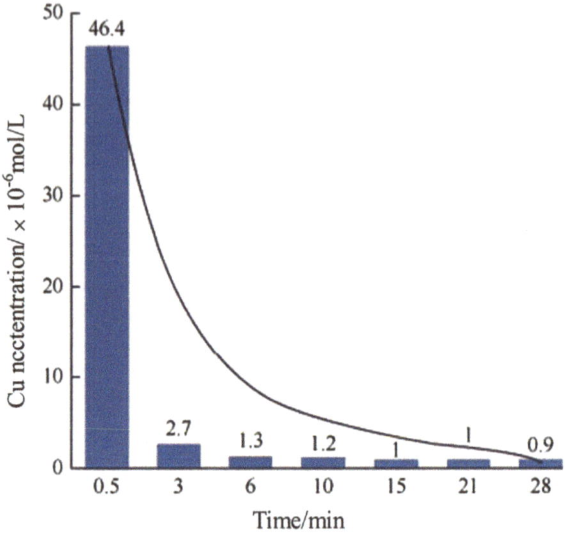

FIGURE 7.2 Curve of concentration and time for copper ions in chalcopyrite aqueous solution.

The XPS analysis (see Fig. 7.3) shows that only Cu, Fe, and S are detected, in addition to C and O, indicating that the minerals used in the test are relatively pure. The peak intensities of the photoelectron spectroscopy of the Cu 2*p* and S 2*p* before and after treatment with copper ions were compared to explain the number change in the Cu and S on the surface before and after adsorption of the samples. Fig. 7.3 shows the XPS spectra of the chalcopyrite after treatment of the crude ore with copper ions. Fig. 7.3C and E shows the energy spectra of the S 2*p* and Cu 2*p* on the chalcopyrite surface, respectively; Fig. 7.3D and F, the respective energy spectra of S 2*p* and Cu 2*p* on the chalcopyrite surface after the interaction with copper ions; and Fig. 7.3A and B, the comprehensive spectra of chalcopyrite samples before and after treatment with copper ions.

Fig. 7.3 shows that the relative peaks of the Cu 2*p* and the S 2*p* are larger after adsorption than before adsorption. Comparison of this change before and after adsorption shows that the concentration of Cu on the surface of the treated chalcopyrite sample increases with the decreased concentration of S. This result indicates that the copper ions are adsorbed on the chalcopyrite surface, interacting with S on the surface. The same results are also obtained in the comparison of the relative intensities of Cu 2*p* and S 2*p* in the comprehensive spectrum.

Table 7.1 shows the semiquantitative analysis of atomic concentration on the surface of the local site of XPS. The results in Table 7.1 show that the atom concentration of Cu on the chalcopyrite surface before and after treatment with copper ions increases from 33.44% to 38.19%, and the atomic concentration of

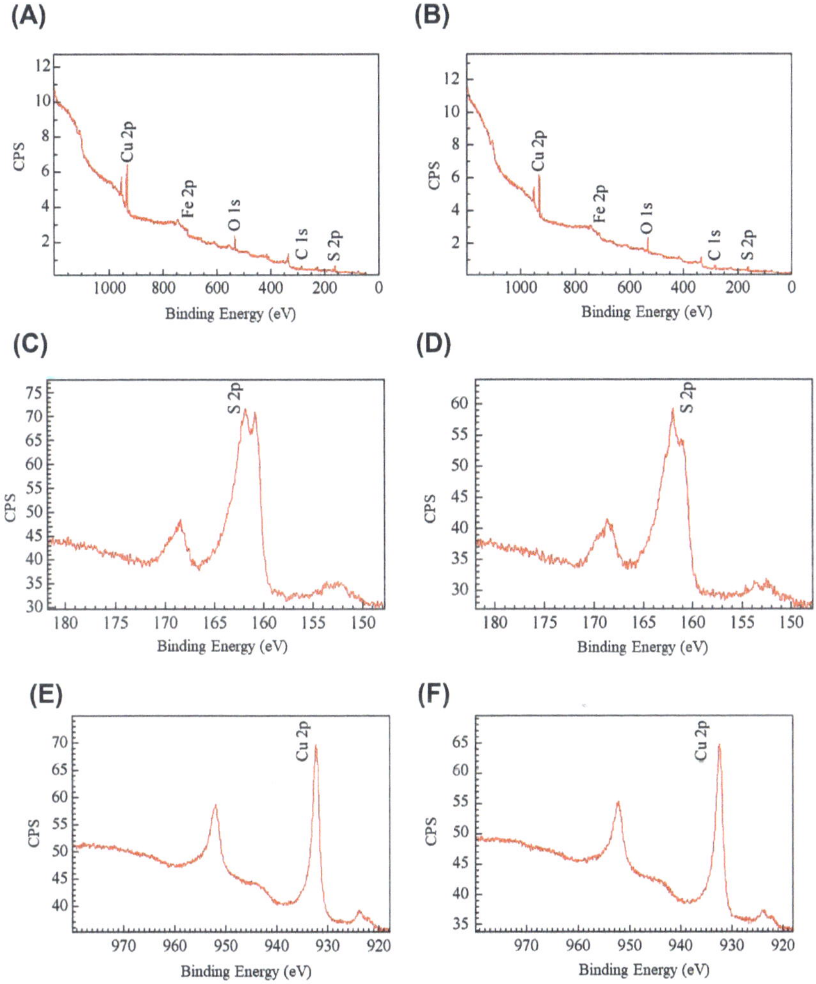

FIGURE 7.3 X-ray photoelectron spectroscopy spectra of the untreated and treated chalcopyrite. (A and B) The comprehensive spectra of chalcopyrite samples before and after treatment with copper ions. (C and E) The energy spectra of the S 2*p* and Cu 2*p* on the chalcopyrite surface, respectively. (D and F) The respective energy spectra of S 2*p* and Cu 2*p* on the chalcopyrite surface after the interaction with copper ions.

S decreases from 45.38% to 42.10%. This indicates that the crude ore and copper ions have surface adsorption, resulting in a relative increase in the atomic concentration of Cu on the surface. In addition, Fig. 7.3 shows that the peak shape of S 2*p* before and after the treatment with copper ions also changed, with the shifted binding energy. The binding energy of the region of variation corresponds to S^{2-} with a Cu–S bond, which confirms that Cu interacts with the sulfur ions on the S-rich surface.

TABLE 7.1 Semiquantitative analysis of atom concentration on the chalcopyrite surface.

	Element	Mass concentration (%)	Atomic concentration (%)
Raw material	S 2*p*	30.55	45.38
	Cu 2*p*	44.57	33.44
	Fe 2*p*	24.87	21.21
Cu^{2+} activation	S 2*p*	27.68	42.10
	Cu 2*p*	49.76	38.19
	Fe 2*p*	22.56	19.71

Based on these results, the metal components released by the inclusions in chalcopyrite, especially Cu, are adsorbed on the chalcopyrite surface and increase the ζ potential.

7.1.3 Surface adsorption of inclusion components in sphalerite

The introduction to Chapter 5 proves that there are a large number of fluid inclusions in sphalerite, which are rich in Zn. The concentration of Zn measured in the fluid inclusions of sphalerite after 10 min of grinding was 18.35×10^{-6} mol/L. Based on theoretical calculation under weak acid conditions and alkaline conditions ($6 \leq pH < 12$), the order of magnitude of concentration of Zn released by sphalerite dissolution was in the range of 10^{-7} to 10^{-8} mol/L. Thus the Zn released by the fluid inclusions of sphalerite at this pH range contributes to the Zn in solution. In other words, there is a large amount of zinc ions released into the pulp solution after the grinding of sphalerite. The Zn released from the fluid inclusions of sphalerite is likely to be adsorbed on the sphalerite surface, thus affecting the electrical properties of the sphalerite surface. The change in the electrical properties has an important influence on the interaction mode of the metal ions and collectors. Therefore, taking the Zn released from the fluid inclusions of sphalerite adsorbed on the sphalerite surface as an example, this section illustrates the interaction between the released components of fluid inclusions and the mineral surfaces.

The IEP of the determined sphalerite in pure water is pH 6.5, which is within the range (2–8) of IEP of sphalerite in the literature. When $pH < 6.5$, the sphalerite surface shows positive electricity; when $pH > 6.5$, the surface shows negative electricity. It takes a certain time for the zinc ions in the solution to be adsorbed on the sphalerite surface. The adsorption process can be characterized by the dynamic relationship between the ζ potential of the

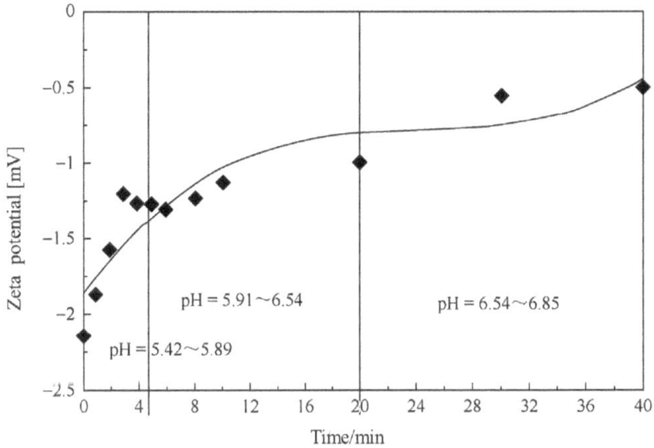

FIGURE 7.4 ζ potential versus time in aqueous solution of sphalerite.

sphalerite surface and time. Fig. 7.4 shows the change in the ζ potential of sphalerite after grinding in the pure water (pH 6.8) with time. From the solution chemistry calculation of Zn^{2+} in Chapter 6, at pH 6.8, most of Zn in the solution exists as Zn^{2+}, and the remaining hydrolyzed components are less than 2%. Therefore, the change in ζ potential on the sphalerite surface at pH 6.8 is due to the interaction of Zn^{2+} with the mineral surfaces, rather than the effect of Zn hydroxide.

Fig. 7.4 shows that the sphalerite surface in pure water is negatively charged. When $t = 0$, the ζ potential of the sphalerite surface is −2.14 mV. As the time increases, the ζ potential gradually increases. In the first 10 min, the potential rises significantly from −2.14 to −1.13 mV, and especially within the first 4 min, the potential increases linearly with time. This indicates that the zinc ions are strongly adsorbed on the sphalerite surface in the first 10 min. After 10 min, the potential slowly rises and reaches equilibrium in about 30 min. The equilibrium potential is about −0.5 mV. The rise in ζ potential on the sphalerite surface with time indicates that Zn released from the fluid inclusions is adsorbed on the sphalerite surface.

To verify the adsorption rule and degree of Zn adsorbed on the sphalerite surface, we measured the total Zn concentration of sphalerite in pure water within the first 30 min of the change in the surface potential test (see Table 7.2).

Table 7.2 shows that the Zn concentration in the solution decreases significantly with time in the first 10 min. This indicates that the zinc ions in the solution are adsorbed on the sphalerite surface; this is consistent with the trend of the potential on the sphalerite surface to increase significantly with time in Fig. 7.3. After 10 min, the Zn in solution increases slowly with time, due to the Zn released by the dissolution of sphalerite. However, compared with the initial Zn concentration ($t = 0$), the concentration of Zn in the

TABLE 7.2 Relationship between concentration of Zn and time in sphalerite aqueous solution.

Time (min)	Zn concentration in solution ($\times 10^{-6}$ mol/L)
0	19.0
2	4.02
5	3.77
10	2.91
15	4.46
20	5.69
25	6.89

solution is reduced within the time range. In summary, the Zn released by the fluid inclusions in sphalerite is adsorbed on the sphalerite surface, thus increasing the surface potential of sphalerite.

7.1.4 Surface adsorption of inclusion components in pyrite and galena

As described in Chapter 5, fluid inclusions in pyrite crystals contain ore-forming ions, chloride salts, and sulfates in the diagenetic fluids captured during mineralization. The concentrations of Cu and Fe in the aqueous solution after the release of the fluid inclusions of pyrite from Weixin were 3.29×10^{-6} and 32.52×10^{-6} mol/L, respectively, while the concentrations of Cu, Pb, Zn, and Fe in the aqueous solution from Dapingzhang were 2.43×10^{-7}, 16.51×10^{-7}, 19.45×10^{-7}, and 516.52×10^{-7} mol/L, respectively. The electrochemical properties of these sulfide minerals had significant influence on their flotation behaviors. These ions change the electrochemical properties of the pyrite surface as well as the flotation behavior. To investigate the effect of the release of the mineral fluid inclusion components on the surface electrical properties of the minerals, the ζ potential of the pyrites from Weixin and Dapingzhang was measured soon after grinding without washing (see Fig. 7.5).

Pyrites from the two regions showed slight oscillations in ζ potential throughout the test period due to the complex reaction caused by the ions in solution. The overall trend of ζ potential showed that in 0–4 min, the ζ potential of pyrite gradually increases. In the pyrite from Weixin there was an increase from 0.14 to about 0.20 mV, and the pH decreased from the initial 6.98 to about 4.5.The rising trend of pyrite from Dapingzhang is more obvious,

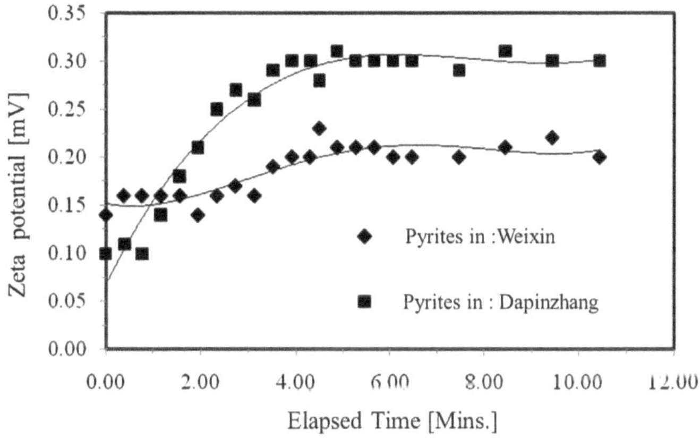

FIGURE 7.5 ζ potential of pyrite in aqueous solution versus time.

from 0.1 to 0.35 mV, and the pH changed from 6.98 to 3.20. Within 4–12 min, the potential and pH tended to equilibrate, indicating that the various reactions in the solution balance themselves. The change rule of the ζ potential illustrates the component release of the fluid inclusions in pyrite, that is, some metal ions are adsorbed on the mineral surface in the ore pulp system.

After galena is broken, the detected surface potential changes with time (see Fig. 7.6). The results show that the surface potential increases with time, and reaches equilibrium in 4 min, forming a stable electric double layer. However, the increase in the potential of galena is not as great as that of chalcopyrite and sphalerite. This is because the lead ions in the inclusion components are more hydrolyzed than the copper and zinc ions to form lead hydroxide precipitation, thereby reducing the concentration of lead ions in the solution. In the first 30 s, the pH of the ore pulp dropped sharply, which proves the hydrolysis of lead ions.

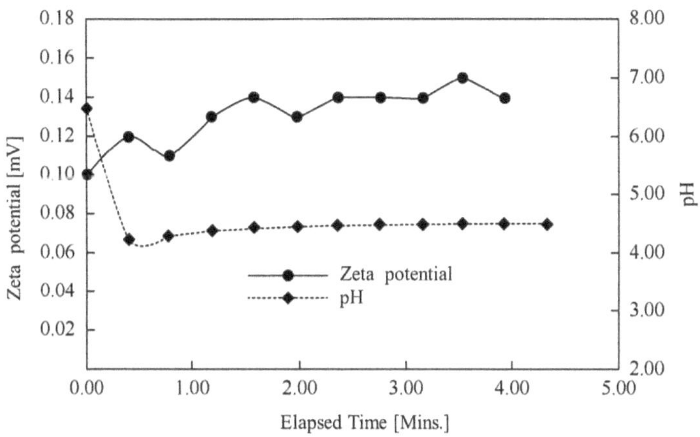

FIGURE 7.6 ζ potential versus time in aqueous solution of galena.

7.2 Density functional theory of the interactions between the components of fluid inclusion colonies and mineral surfaces

7.2.1 Crystal model and calculation method

With respect to the DFT, two common commercial software packages, Materials Studio (MS) and Vienna Ab Initio Simulation Package (VASP), were used in this work. MS is the specialized software for modeling and simulation developed by Accelrys, and is widely used in fields such as chemistry, materials, and biology. Supporting a variety of operating platforms such as Windows and Linux, it is convenient to build a three-dimensional structural model and study crystals, amorphous and polymer materials, and the related processes. Reliable data can be obtained in configuration optimization, property prediction, and X-ray diffraction analysis, as well as complex dynamics simulations and quantum mechanical calculations. VASP is a software package for calculation in ab initio quantum mechanics and molecular dynamics using pseudopotential and plane wave basis sets. The method in VASP, based on the local density approximation at finite temperature, uses the diagonal matrix and Pulay mixing for instantaneous electronic ground state. These techniques can avoid problems in the original Car–Parrinello method. The interaction of ions and electrons is described by the Vanderbilt pseudopotential or projector augmented wave (PAW). VASP uses periodic boundary conditions (or supercell models) to handle the properties of atoms, molecules, clusters, crystals, and amorphous materials, as well as surface system relaxation, reconstruction, scanning tunneling microscopy simulation, and geometry and electronic structure calculations.

The general idea is as follows: first, the mineral single crystal cell is constructed and geometrically optimized. Then, a supercell surface model of suitable size is constructed on the basis of the optimized single crystal cell to study the geometric optimization and supercellular surface properties. Subsequent effects with other metal component collectors are performed on the optimized supercell surface model.

After the release of the metal components in the inclusions of sulfide mineral, especially the heavy metal activation components on the surface, migration and adsorption occur on the mineral surfaces. Due to the large number of metal components in the inclusions, Cu, the typically activated component in the inclusions of sulfide mineral, is selected for research. There are two forms of copper ions in sulfide minerals: replacing the metal atoms on the mineral surfaces and being adsorbed on the mineral surfaces. The interaction energy in the calculation system is an important basis for judging whether the interaction occurs. The interaction between the sulfide mineral surfaces and Cu can be calculated as follows.

Substitution effect:

$$\Delta E_{sub} = E^{tot}_{slab+Cu} + E_{Cu} - E^{tot}_{slab} - E_M \tag{7.5}$$

where E^{tot}_{slab} and $E^{tot}_{slab+Cu}$ represent the energies of the calculated mineral surface structure model before and after the Cu substitution, respectively; E_{Cu} and E_M represent the energies of Cu and the substituted atom, respectively; and ΔE_{sub} represents the substitution energy of Cu substitution on M, the metal atom on the mineral surface.

Adsorption:

$$\Delta E_{ads} = E^{tot}_{slab+Cu} - E_{Cu} - E^{tot}_{slab} \tag{7.6}$$

where E^{tot}_{slab} and $E^{tot}_{slab+Cu}$ represent the total energies of the calculated mineral surface model before and after Cu adsorption, respectively; E_{Cu} represents the energy of Cu; and ΔE_{ads} represents the adsorption energy of Cu on the mineral surface.

The interaction energy ΔE_{int} between the sulfide mineral surface and the collector (xanthate) is calculated as:

$$\Delta E_{int} = E^{tot}_{EX+slab} - E_{EX} - E^{tot}_{slab} \tag{7.7}$$

where E^{tot}_{slab} and $E^{tot}_{EX+slab}$ represent the total energies before and after the interaction between the system and xanthate, respectively; E_{EX} represents the energy of xanthate; and ΔE_{int} represents the interaction energy of xanthate molecule on the mineral surface.

It should be noted that since the calculation of the DFT simulation is performed in a vacuum, the calculation function can be different depending on the calculation system. Therefore, they do not represent real energy, and the energy calculated by DFT provides only a qualitative basis for the reaction system. That is, if the calculated interaction energy is negative, the interaction will occur. A highly negative value indicates that adsorption is more likely to occur. A positive energy value means that the reaction cannot proceed.

7.2.1.1 Chalcopyrite crystal model and calculation method

Taking into account the periodic boundary conditions, the unit cell and supercell models of the chalcopyrite crystal were calculated using the Cambridge Serial Total Energy Package (CASTEP) in MS. The electron–ion interaction was detected using the optimized ultrasoft pseudopotential of Vanderbilt. The generalized gradient approximation (GGA) was used for calculation, and Perdew–Wang (PW91) was used for exchanging correlation energy. The structural optimization considered the Broyden–Fletcher–Goldfarb–Shanno (BFGS) algorithm, and the force on the atoms was set to be less than 0.1 eV/nm to calculate the convergence accuracy. All calculations were in reciprocal space, and the spin state of Fe was set. The truncation of the plane wave affected the results, and was set to 500 eV.

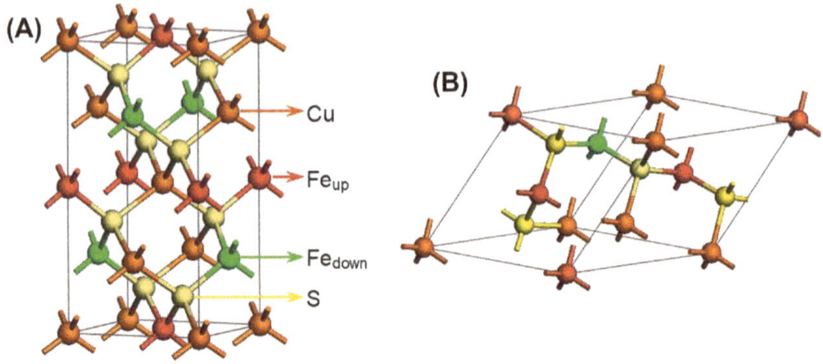

FIGURE 7.7 Chalcopyrite crystal structure. (A) Supercell. (B) Unit cell.

The unit cell of chalcopyrite contains 4 Cu, 4 Fe, and 8 S (see Fig. 7.7A). Its structure can be regarded as sphalerite with a tetragonal system, where half of the cations are replaced by copper ions and the other half by iron ions. This causes a slight deformation in the *z*-axis direction, and c/a = 1.971. The supercell structure $Cu_4Fe_4S_8$ is converted to the original cell, $Cu_2Fe_2S_4$ (see Fig. 7.7B).

On the basis of the aforementioned unit cell optimization, a 2 × 2 × 2 chalcopyrite (0 0 1) surface supercell model was constructed, and the surface atomic layer was a metal layer. VASP was used to calculate the supercell model of the chalcopyrite (0 0 1) surface and its subsequent interaction with other components. We used the Perdew–Burke–Ernzerhof exchange (PBE) in GGA for the Schrodinger equation and PAW to describe the electron–ion interaction. We developed the valence electron Kohn–Sham wave function with the plane wave basis set, and used the Monkhorst–Pack grid to divide the K-point grid into the irreducible part of the Brillouin zone. Then we applied the linear tetrahedron modification of Blöchl for the accurate calculation system, and optimized the ground-state atomic configuration using the Hellmann–Feynman force field. During the relaxation, the Hellmann–Feynman force of each atom was calculated after the electron wave function was self-consistent, and the magnitude and direction of the force were used to adjust the position of each atom. We set the H–F force on each atom to be less than 0.0005 eV/nm.

The calculations were performed on a minimized fast Fourier transform grid. For the surface structure of chalcopyrite (0 0 1), a 2 × 2 × 2 supercell was selected in directions a and b, and eight atomic layers were in direction c. To avoid interlayer interaction, vacuum filling was used, with the vacuum thickness of 15 Å. After the necessary numerical tests, this structural model was sufficient to ensure the calculation. In the calculation, an adsorbate molecule (Cu or xanthate molecule) was introduced into the optimized structure. Before and after the introduction, the K-point mesh of the Brillouin zone was 3 × 3 × 1, and the dynamic truncation energy was selected as 500 eV. The convergence test was performed by changing the sampling point density and the cutoff energy of the K space, and the results showed that these settings were sufficient for the accuracy of the calculation.

7.2.1.2 Sphalerite crystal model and calculation method

The sphalerite quantum chemical calculation was performed by CASTEP in MS, which adopts the first principle under the DFT. The exchange-correlation function in the calculation was described by the PBE gradient correction function under GGA. The ultrasoft pseudopotential (Ultrosoft) of the PW91 basis set was used to describe the interaction between the ion cores and the valence electrons. The pseudopotential calculation of each atom selected the valence electron structure of the corresponding atom based on the periodic table of chemical elements. For example, the valence electrons of Zn and S were Zn $3d^{10}4s^2$ and S $3s^23p^4$, respectively. The plane wave truncation was set to 285 eV, and the total energy and charge density of the system in the Brillouin area was calculated using the Monkhorst–Pack. The k-grid point 2 × 3 × 1 was selected to ensure the convergence of the system's energy and the configuration at the quasi-complete plane wave base.

In the self-consistent field operation, Pulay density mixing was adopted, with the convergence precision set to 1.22×10^{-6} eV·atom^{-1}. The BFGS algorithm was used in the model structure optimization. The convergence criterion of the optimization parameters, including the interaction force between the atoms, was set to 0.005 eV nm^{-1}; the convergence criterion of the internal stress of the crystal was set to 0.1 GPa; the maximum displacement of the atom was set to 0.002 Å. Fig. 7.8 shows the unit cell model (Zn_4S_4) and the

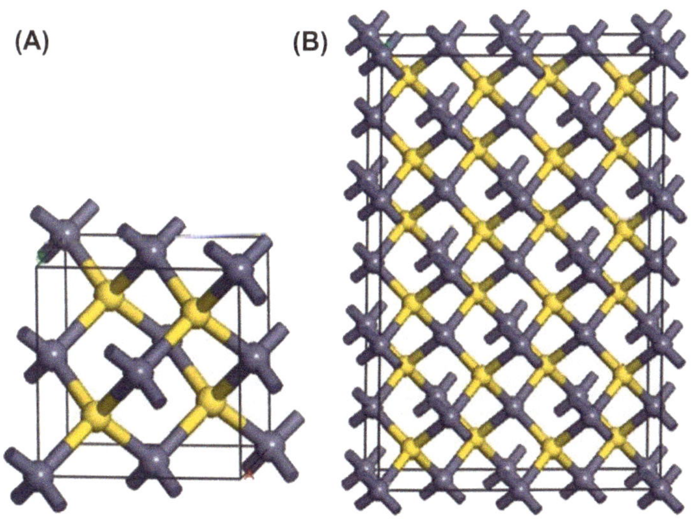

FIGURE 7.8 Unit and supercell model of ZnS. (A) Unit cell. (B) Supercell.

$2 \times 1 \times 3$ supercell model ($Zn_{24}S_{24}$) of sphalerite. Fig. 7.8A shows that in the sphalerite crystal, each single crystal cell contains 4 Zn and 4 S. One-fourth of the diagonals are S, and the eight corners and six centroids are Zn. The sulfur ions in the crystal are closely packed in the cubic, and the zinc ions are located in the tetrahedral voids with the coordination number of 4.

Based on the above single-cell model, a supercell model of $2 \times 1 \times 3$ ZnS (110) surface was constructed for subsequent calculation of the interactions with Cu and the collectors. The supercell model on the ZnS (110) surface contained six atomic layers with the vacuum layer thickness of 10 Å. When the calculation system involved the organic xanthate molecule, the DMol3 in MS was used to optimize the structure of the xanthate molecule. The preliminary optimized xanthate molecular model was submitted to a periodic box of the same size as the ZnS supercell model for optimization, and then the structurally optimized xanthate molecule and the ZnS supercell model were applied. The adsorption calculation of the collector molecules was set to 351 eV. All calculations were performed under the medium grid, ignoring the spin polarization of the system in the calculation.

7.2.1.3 Pyrite crystal model and calculation method

First, the theoretical unit cell (see Fig. 7.9A) and the $2 \times 2 \times 2$ supercell model (see Fig. 7.9B) of the ideal crystalline pyrite were constructed using CASTEP in MS. The exchange-correlation energy in the calculation was described by PBE in GGA, and the K point was sampled by the Monkhorst-Pack. To ensure the reliability of the results, the truncation energy test was performed before calculating the properties. The fixed K-points grid was $2 \times 2 \times 2$, with the truncation energy of the final plane wave of 330 eV. The atomic electrons were ultrasoft pseudopotential and calculated in the reciprocal space of the crystal. The convergence accuracy of the structurally optimized electron energy eigenvalue was 2.0×10^{-6} eV, with average atomic

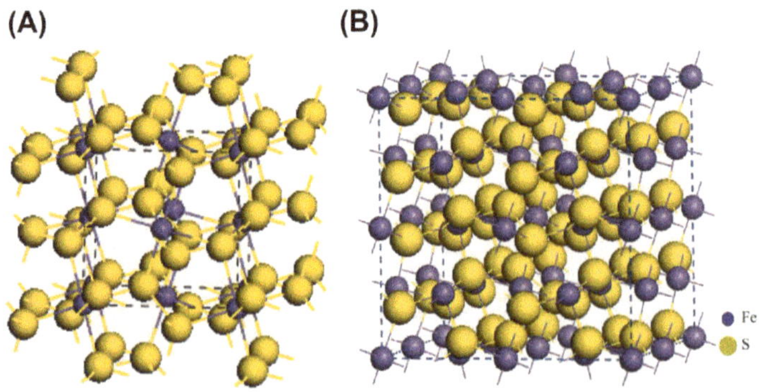

FIGURE 7.9 Theoretical model of pyrite bulk. (A) Unit cell structure. (B) $2 \times 2 \times 2$ supercell structure in the region of *blue dotted line*).

force not more than 0.05 eV/Å. The electrons were self-consistent with the Pulay density mixing. The same parameters as in the geometric optimization were used to calculate all properties, and the smearing value was 0.1 eV when the density of states was calculated.

The theoretical model showed that each unit cell of pyrite contained four FeS_2 molecular units, with the Fe distributed in the six centroids and eight vertex angles of the cubic unit cell. Each Fe was coordinated with six adjacent S to form an octahedral structure; each S was coordinated with three Fe and one S to form a tetrahedral structure. A dumbbell-like structure was formed between the two S in the form of a sulfur dimer (S2-2), and was arranged along the (111) surface.

On this basis, the supercell models of pyrite (110) and (311) surfaces were constructed for subsequent correlation properties and calculations with other components. The calculation of the correlation function and the convergence accuracy were similar to the unit cell calculation, with the truncation energy set to 300 eV. In addition, the surface supercell model was tested for different atomic layer thicknesses and vacuum layer thicknesses, thus ensuring that the properties of the near-middle and lower layers of atoms were consistent with those in the bulk phase. The results showed that the surface energy changes were small when there were 12 layers of atoms in the surface model. The test of the vacuum layer thickness showed that the surface energy with the vacuum layer thickness of 10 Å was the lowest. Therefore, the surface structure of 12 layers of atoms and the vacuum layer thickness of 10 Å showed satisfactory convergence.

7.2.2 Surface relaxation and reconstitution of sulfide mineral

Cu–Pb–Zn sulfide mineral is a typical crystalline mineral. During crushing, the particle size of the mineral gradually reduces to form new surfaces. The arrangement of atoms on the surface of the newborn is related to the internal geometry of the crystal, significantly different from the internal structure of the crystal. The difference is caused by the relaxation and reconstruction of the surface atoms during surface formation. The relaxation of the atoms on the crystal surface is manifested by the displacement of the surface atoms perpendicular to the surface. The reason for the relaxation is that when the new surface is formed, the stress on the surface atoms is asymmetrical, with high surface energy and unstable surface structure. To stabilize the surface, the surface atoms are displaced in the direction that reduces the surface energy.

The relaxation and reconstitution of the surface atoms of metal sulfide mineral changes the geometry of the surface atoms, eventually leading to changes in the surface properties. The relaxation and reconstitution on the newly formed surface of the sulfide mineral have an important influence on the adsorption of the inclusion components on its surface, as well as the interaction between the flotation agents. To a large extent, the "innate" surface properties of minerals, such as the arrangement geometry of the surface atom, determine the behaviors of the metal ions and flotation agents on the mineral surface. Therefore, before studying the inclusion components and minerals, the surface structure of the sulfide mineral should be studied. This is important for understanding the microscopic nature of the mineralized surface and the mechanism of interactions among the surface of the sulfide mineral, inclusions, and collector agents.

7.2.2.1 Surface structure of chalcopyrite

The chalcopyrite (0 0 1) surface has two forms: one is the metal end, i.e., the M surface; the other is the S end, i.e., the S surface. The S end shows the S-rich surface after relaxation. Significant relaxation appears on both surfaces. The M surface is a parallel zigzag chain composed of copper ions and iron ions, and the S surface is composed of sulfur ions forming the reverse bond in the bottom layer. After relaxation, the chalcopyrite dissociates along the (0 0 1) surface, and then undergoes relaxation to form an electrically neutral surface. The surface relaxation causes the following characteristics: The atoms in the body positions are displaced, resulting in small rotations and folding (Δd) between the adjacent metal and the S layers (see Fig. 7.10).

Fig. 7.11 shows the (0 0 1) surface of the metal end: Fig. 7.11A is the model before geometry optimization, and Fig. 7.11B is the geometry-optimized model. Fig. 7.11A shows the arrangement of S, Cu, and Fe in the different layers of the chalcopyrite (0 0 1) surface; Fig. 7.11B shows the structure after the energy convergence. The atomic arrangement of the chalcopyrite (0 0 1) surface has no preoptimization rules, hence the arrangement of S, Cu, and Fe is slightly changed. The change in the S is quite obvious. After the energy convergence tends to stabilize, the S is displaced outward in the *z*-axis direction, where the cell expansion occurs. Before and after the relaxation, the lengths of S–Fe and

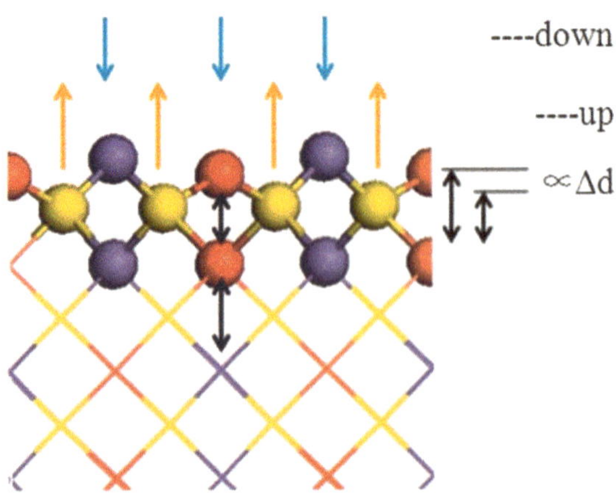

FIGURE 7.10 Surface relaxation and fold schematic diagram of chalcopyrite.

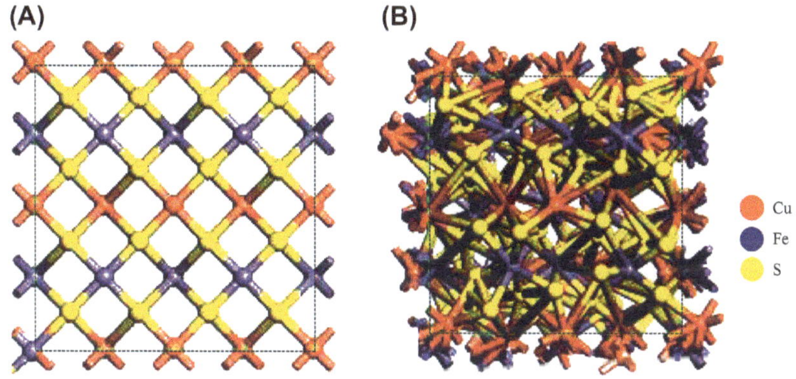

FIGURE 7.11 Chalcopyrite (0 0 1) surface. (A) Before optimization and (B) after optimization.

S—Cu bonds change greatly. These cause the crystalline S to expand outward. The atomic arrangement is also changed in the *x*- and *y*-axis directions. The value of the *z*-axis direction increases in the optimized model, indicating that cell expansion has occurred. The volume of the unit cell becomes larger in the crystal structure. This is because the crystal structure tends to be stable, and the S has strong interactions with the neighboring atoms. During the structural optimization, the surface relaxation and structural reconstruction occur on the (0 0 1) surface of the crystal structure.

Furthermore, in the calculation model, the first metal layer of the metal end surface without relaxation is the Cu—Fe metal atomic layer; S is in the second layer; the third layer is the metal atomic layer, and so on. After relaxation, the atoms of the first layer fall, while those of the third metal layer rise, forming a special structure of metal layers. The model of the metal atom formation includes Fe—Fe, Fe—Cu, and Cu—Cu bonds, and S in the second metal layer on the (0 0 1) M surface occupies these regions by up-and-down alternation. Each region unit has Fe—Fe, Cu—Cu, and Fe—Cu bonds, but all the bonds have similar bond lengths and form a periodic structure. Each structural unit has a layer of S located on the upper and lower metal surface layers, and the changes in the lengths of the S—Fe and S—Cu bonds after relaxation are different, especially Cu—S bonds.

A similar reconstitution after the mineral dissociation obtained by plane wave in other reports can be explained by the dangling bonds. In the chalcopyrite, each metal atom links and coordinates four S. In this case, the *dxy*, *dyz*, and *dxz* orbits of both Fe and Cu overlap with the *sp*3 orbit. However, the *dxy*, *dyz*, and *dxz* orbits do not overlap with the S orbit on the surface after dissociation, thus forming the dangling bonds. To balance the dangling bonds, the metal atoms on the surface fall, while the metal atoms of the third atomic layer rise outward. In the same plane layer, the *dxy*, *dyz*, and *dxz* orbits overlap to form the δ bond between the metal atoms, which is consistent with the bond length of the surface metal. This also causes the upper and lower S to have an upward and downward tendency with respect to the metal layers of Fe and Cu. These are the mechanisms formed by the relaxation and reconstitution of the chalcopyrite (0 0 1) surface.

To verify the theoretical calculation results and study the surface structure of chalcopyrite, we performed a microscopic analysis of the atomic force on the crystal surface. A large number of pure crystals of chalcopyrite were cut to obtain a series of fresh surfaces. Then the smooth surfaces were selected using atomic force microscopy (AFM) to determine the direction of the surface atoms and the interatomic distance, thus obtaining the corresponding dissociation surfaces. Different positions of the same sample were scanned to ensure a certain breadth of the same dissociation surface. There were dissociation surfaces in chalcopyrite. In this section, the chalcopyrite with the (0 0 1) surface as the dissociation surface were selected for AFM characterization.

The chalcopyrite crystals were broken to form surfaces, which were scanned by AFM to obtain the surface topology images at the atomic scale (see Figs. 7.12–7. 16). It is topology because when observed from the microscopic angle at the atomic scale, the surface is moving and changing. The images scanned at different times are different, but the surface substance reflected by them is unchanged.

Fig. 7.12 shows that the chalcopyrite surface is extremely uneven at the atomic scale. From a more microscopic point of view, there is a difference in each place, indicating that the atomic arrangement on the chalcopyrite surface

FIGURE 7.12 Atomic force microscopy three-dimensional image of the chalcopyrite crystal.

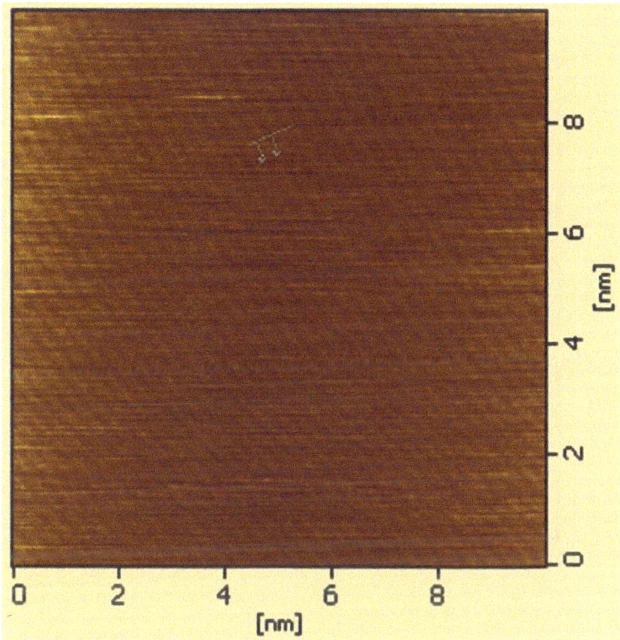

FIGURE 7.13 Ichnograph of the chalcopyrite surface (site a).

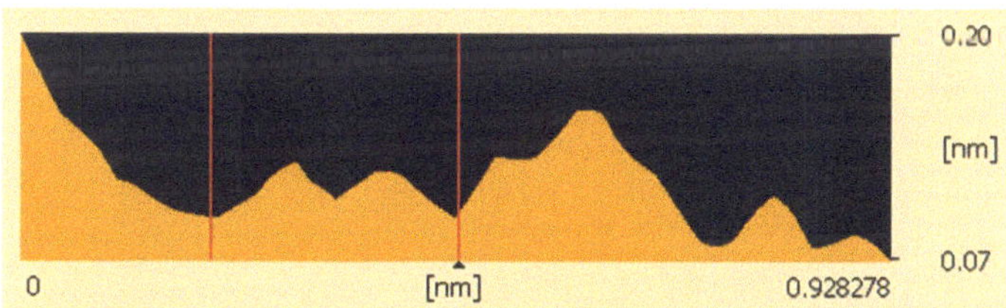

	Z1[nm]	Z2[nm]	ΔZ [nm]	Distance [nm]	Φ[°]
■	0.098347	0.100910	0.002563	0.264769	0.554638

FIGURE 7.14 Two-dimensional topological graph of atoms on the chalcopyrite surface (site a).

is irregular compared with the regular arrangement of atoms in the crystals. Comparing the atomic radii of Cu^{2+}, Fe^{2+}, and S^{2-}, the interatomic distance is smaller than the superposition of the isolated ion diameters after the bonding of Cu^{2+}, Fe^{2+}, and S^{2-}. If the chalcopyrite surface is ordered in a ratio of 1:1:2 for Cu, Fe, and S, the region in Fig. 7.12 will contain more than 376 atoms.

In the spatial direction, the heights of the surface atoms are inconsistent with the height difference of 0.3 nm, which is close to the diameter of a sulfur ion. That is to say, on the standard chalcopyrite surface, there is an uneven atomic order of magnitude. Figs. 7.14 and 7.16 show that the two-dimensional electron distribution of atoms on the chalcopyrite surface has a height

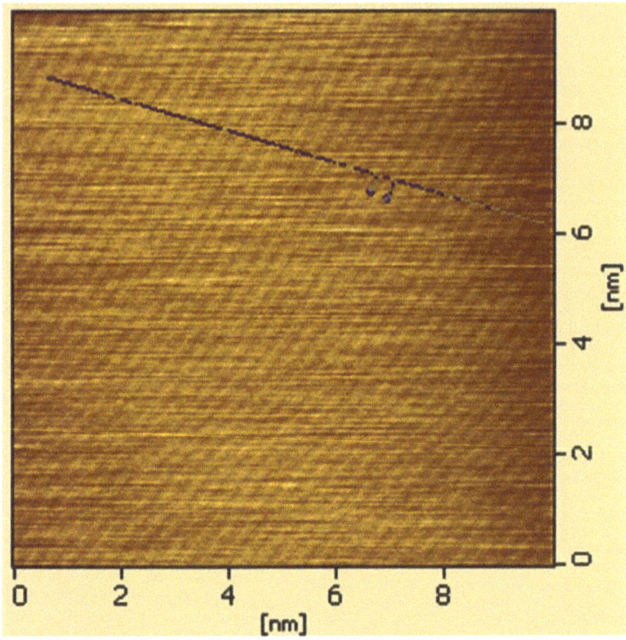

FIGURE 7.15 Ichnograph of the chalcopyrite surface (site b).

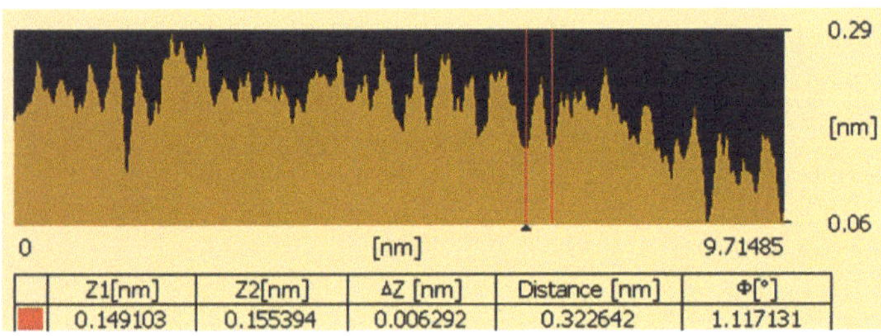

FIGURE 7.16 Two-dimensional topological graph of atoms on the chalcopyrite surface (site b).

difference of 0.2–0.29 nm in the spatial direction, which indicates that the atoms on the chalcopyrite surface are not smooth in the spatial direction. That is, some surface atoms are displaced in the spatial direction relative to their original positions, namely surface relaxation.

During the formation of the chalcopyrite surface, the cracked chemical bonds are unstable, with high energy. To reduce the surface energy, the cracked chemical bonds will adsorb the atoms of foreign substances. Even in a vacuum, the chemical bonds breaking at the surfaces reduce their energy by bonding. Therefore, the positions of the surface atoms are changed relative to their original regular positions, and this surface reconstruction is due to the random, irregular arrangement.

The AFM atomic scale of the chalcopyrite surface in Fig. 7.12 shows that the uneven surface atoms are not evenly distributed like the uniform distribution of the interior of the crystal. The two-dimensional distribution of the electron clouds of multiple atoms on the chalcopyrite surface in Fig. 7.16 shows that the distance between the atoms is not regular in the horizontal direction. From the scale of the atomic aggregates, some atoms overlap. In Figs. 7.13 and 7.15, the atomic arrangement on the chalcopyrite surface is obviously stripe shaped—the stripes with higher brightness are in the protuberant atomic arrangement; those with lower brightness are in the cupped atomic arrangement. The atoms are also arranged in parallel along a 45-degree angle in the figure. The region between the two vertical lines in Fig. 7.14 is the two-dimensional electron cloud distribution of the protuberant atoms in Fig. 7.13. The diameter of the atom is 0.265 nm, which is between the diameters of S^{2-} and O^{2-} ions, and comes closer to the diameter of S^{2-} after losing some negative charge. Therefore, it is S bonded with copper or iron ions. The region between the two red lines in Fig. 7.16 is the protuberant atoms in Fig. 7.15, with an atomic diameter of 0.323 nm, which is close to the diameter of S^{2-} after losing some negative charge. The test results show that the relaxation of the chalcopyrite surface brings S to the outer region of the surface, which is consistent with the DFT calculation.

7.2.2.2 Surface structure of sphalerite

The relaxation of the atoms on the crystal surface is manifested by the displacement of the surface atoms in the normal direction perpendicular to the surface. The reason for the relaxation is that when the new surface is formed, the stress on the surface atoms becomes asymmetrical, with high surface energy, and unstable in structure. To stabilize the surface, the surface atoms are displaced in the normal direction, thus reducing the surface energy. To study the atomic relaxation of the sphalerite surface, the sphalerite (110) surface was optimized. Fig. 7.17 shows the surface model before and after the

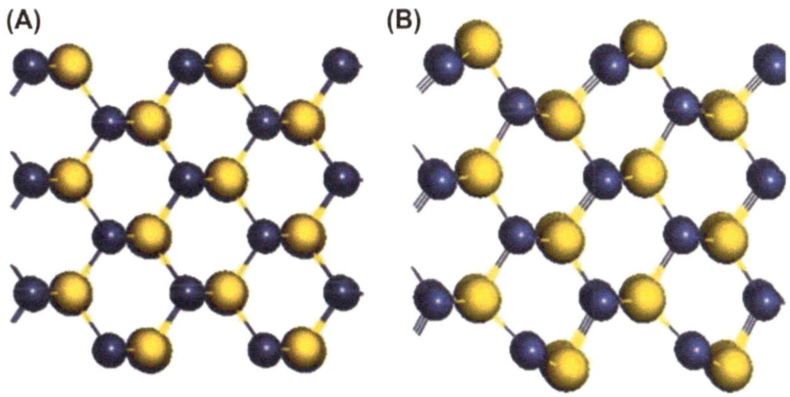

FIGURE 7.17 Sphalerite (ZnS) supercell model of the (110) surface. (A) Before and (B) after geometry optimization.

optimization. The structural optimization is actually the system energy reduction, and surface energy is widely used to measure the stability of the surface atoms relative to the bulk-phase atoms. Its formula is as follows:

$$E_{sur} = \frac{E_{slab} - nE_{bulk}}{2A} \tag{7.8}$$

where E_{sur} is the surface energy per unit area; E_{slab} is the total free energy of the slab; n is the total number of atoms in the slab; E_{bulk} is the free energy of each atom in the crystal; A is the surface area; and 1/2 is the two free surfaces of the simulated unit cells.

As shown in Fig. 7.17A, before the structural optimization of the sphalerite (110) surface, the surface atomic arrangement is basically the same as that of the crystal interior, with regularity. After optimizing the surface structure (the relaxation of surface atoms), the atomic arrangement on the sphalerite (110) surface becomes significantly different from the interior of the crystal. The surface is clearly deformed in geometry (see Fig. 7.17B).

As shown in Fig. 7.18, the sphalerite (110) surface has a significant relaxation. In the direction perpendicular to the sphalerite surface, S in the first layer on the (110) surface is displaced outward by the distance of d_1, and Zn is displaced inside by the distance of d_2. Eventually the S on the sphalerite surface is farthest and forms a relatively S-rich surface. Table 7.3 shows that the displacement distance d_1 of the first layer on the sphalerite surface

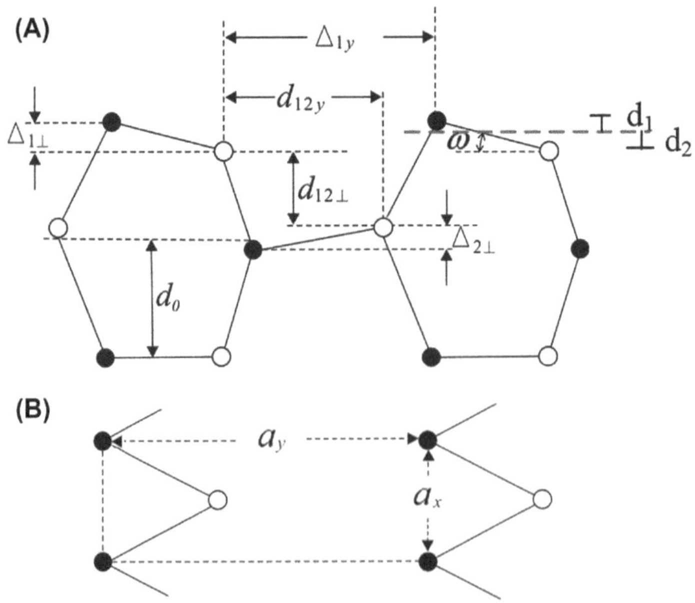

FIGURE 7.18 Schematic plot of the surface geometry for the sphalerite ZnS (110) surface after relaxation. (A) Side view and (B) top view.

TABLE 7.3 Geometric parameters for sphalerite (110) surface after relaxation.

Name	a_y	a_x	$\Delta_{1\perp}$	Δ_{1y}	d_{12y}	$d_{12\perp}$	$\Delta_{2\perp}$	d_0	ω	d_1	d_2
DFT	5.38	3.71	0.56	4.22	3.06	1.37	0.21	1.90	27.39	0.12	0.44
Experimental value	5.41	3.83	0.59	4.29	3.14	1.40	0.00	1.91	28.00	—	—

The unit of ω is °, and that of the rest is Å. *DFT*, density functional theory.

calculated by DFT is 0.12 Å; d_2 is 0.44 Å; the total displacement ($\Delta_{1\perp}$) is 0.56 Å, which is close to the experimental value of 0.59 Å. The unit cell parameter (a_y) of the sphalerite calculated by DFT is 5.38 Å, which is slightly less than the experimental value of 5.41 Å. In addition, in Table 7.3, except for the displacement ($\Delta_{2\perp}$) error of the atoms in the second layer, the results are close to the experimental values, which reflects the reliability of the DFT.

In addition, the data in Table 7.3 show that the displacement of the atoms in each layer on the crystal surface is different. The closer it is to the outermost layer, the more significant the change, that is, $\Delta_{1\perp}$ is significantly larger than $\Delta_{2\perp}$. Eq. (7.8) shows that when the surface energy before relaxation of the atoms on the sphalerite (110) surface is 0.96 J/m^2, the optimized surface energy is reduced to 0.31 J/m^2. This further proved that the atomic relaxation of the sphalerite surface is actually the self-reduction of the surface energy, and the surface structure after relaxation is the most stable.

The sphalerite was cut and polished to form the surfaces, which were scanned by AFM to obtain the atomic-scale surface topological images (see Figs. 7.19 and 7.20).

Fig. 7.19 shows that the sphalerite surface is extremely uneven on the atomic scale. From the microperspective, there are differences everywhere. It shows that the atomic arrangement of the sphalerite surface is completely irregular compared with the regular atomic arrangement in the crystal. The atomic radii of Zn^{2+} and S^{2-} are 0.074 and 0.184 nm, respectively. After Zn and S are bonded, the atomic spacing should be smaller than the superposition of the diameters of the isolated ions. If the sphalerite surface is ordered in a ratio of 1:1 for Zn and S, the total number of Zn and S in the region should be greater than 202 (see Fig. 7.19). In the spatial normal direction, the heights of the surface atoms are inconsistent because there is a current difference of

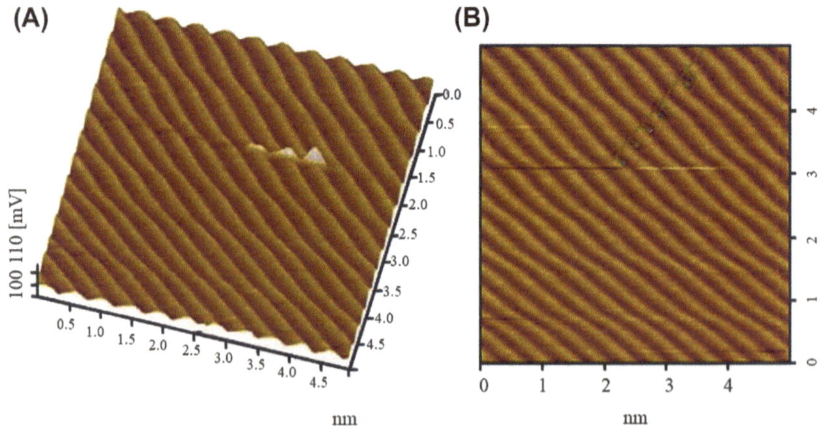

FIGURE 7.19 Atomic force microscopy images of sphalerite surface at the atomic scale. (A) Tridimensional view. (B) Plane view.

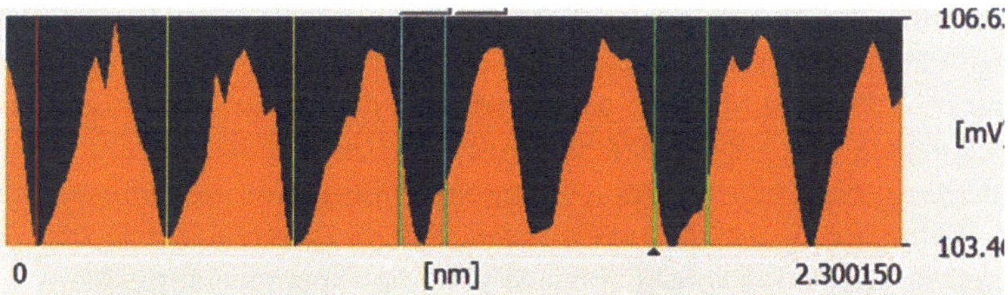

	Z1[mV]	Z2[mV]	ΔZ [mV]	Distance [nm]	Φ[°]
	103.4901	103.6044	0.114318	0.337018	
	103.6044	103.5370	0.067469	0.323538	
	104.9669	104.3411	0.625862	0.112339	
	104.7294	104.3476	0.381798	0.134807	

FIGURE 7.20 Two-dimensional electron cloud topology image of atoms on the sphalerite surface.

10 mV in the space (see Fig. 7.19A). When the probe of the atomic force microscope approaches the surface atoms, the latter electrically repel the probe. Different approaching distances cause different currents, so the currents can indirectly reflect the heights of the surface atoms.

In addition, the sphalerite surface is not even, and its distribution is not as uniform as that in the crystal. Some atoms are almost overlapping, and some are far apart in the horizontal direction, which indicates that the relaxation generated for the sphalerite surface is accompanied by a slight reconstitution. Surface reconstitution means that the base vectors of the two-dimensional unit cells are expanded by an integer multiple. However, because of foreign substances (such as oxygen and carbon dioxide) and relaxation, the integer-multiple expansion of vectors of the two-dimensional unit cells is also affected, thus causing irregular changes.

To determine the arrangement of Zn and S on the sphalerite surface, the distance between the atoms in the positions (see Fig. 7.19B) was measured to obtain the two-dimensional electron cloud topology image of atoms (see Fig. 7.20).

Fig. 7.20 shows that the two-dimensional electron distribution of atoms on the sphalerite surface has certain height differences in the spatial direction, which indicates unevenness of the sphalerite surface in the spatial direction. That is, some surface atoms are displaced in the spatial normal direction relative to their original positions, namely surface relaxation. Herein two atoms are randomly selected in the convex, striped region in Fig. 7.19B to measure the atomic diameter (the atoms between the two red vertical lines and those between the two yellow vertical lines in Fig. 7.20). The atomic diameters are 0.337 and 0.324 nm, respectively. This is close to the diameter of S^{2-} (0.368 nm), and closer to the diameter of S^{2-} after losing some negative charge, which indicates that the atoms in the convex, striped region are S.

Two atoms are also randomly selected in the concave, striped region in Fig. 7.19B to measure the atomic diameter (the atoms between the two blue vertical lines and those between the two green vertical lines in Fig. 7.20). The atomic diameters are 0.112 and 0.135 nm, respectively. This is close to the diameter of Zn^{2+} (0.148 nm), which indicates that the atoms in the concave, striped region are Zn. Therefore, the relaxation and reconstitution of the sphalerite surface indicate that S is the outermost region, which is consistent with the relatively S-rich surface calculated by quantum chemistry. These theoretical calculation results coincide with the experimental results.

7.2.2.3 Surface structure of pyrite

In the crushing of pyrite, the ideal atomic arrangement of the formed surface should be the same as in the bulk phase. However, due to the breakage of the chemical bonds, the chemical bond force of the atoms in the surface layer will be out of balance, and the force on the surface atoms changes. To reach equilibrium, relaxation occurs, and the relaxation of the (100) and (311) surfaces of pyrite is examined below. The theoretical model size of the pyrite (100) surface is as follows: $a_0 = 10.832$ Å, $b_0 = 10.832$ Å, $c_0 = 20.209$ Å. The geometry is rectangular (see Fig. 7.21), and Table 7.4 shows the displacement vector of the surface atoms.

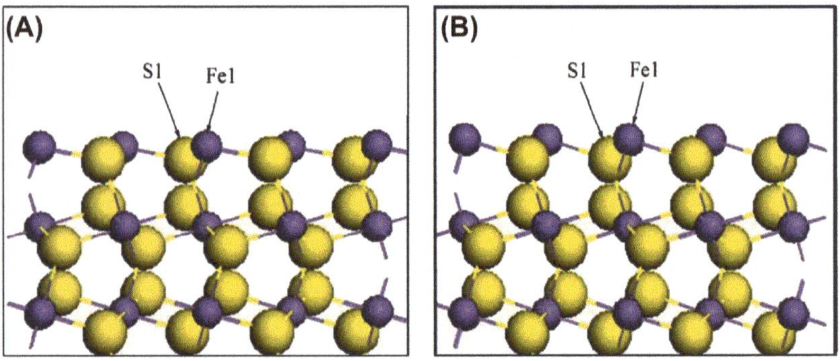

FIGURE 7.21 Pyrite FeS_2 (100) surface. (A) before relaxation. (B) After relaxation.

TABLE 7.4 Pyrite (100) surface atomic coordination and displacements.

Atom label	Coordination number	Charge number (e)	Atomic displacement (Å) Δ_x [0 0 1]	Δ_y [0 1 0]	Δ_z [1 0 0]
Fe1	3	0.14	−0.20	0.26	0.05
S1	3	−0.15	−0.02	0.11	−0.10
Bulk Fe	6	−0.10	/	/	/
Bulk S	4	0.08	/	/	/

Fig. 7.21 shows that the surface cleavage of pyrite reduces the coordination number of the surface atoms. The surface atoms relax without binding of the surrounding atoms. The relaxation of the 3-coordinate Fe in the outermost layer of the (100) surface is most obvious along the Fe1–S1 bond axis of the surface and inner. S in the second layer relaxes upward and outward by 0.11 Å and horizontally toward the coordinated Fe by 0.1 Å. The lateral and horizontal relaxation of the surface atoms of pyrite causes the pyramid of pyrite to shrink inward. Since the atoms below the third layer are all saturated, the relaxation is extremely weak and can be ignored. The mineral relaxation is essentially energy reduction, which is confirmed by the change in the surface energy before and after the surface optimization. The relaxation of the pyrite surface is mainly based on Fe with insufficient surface coordination, which "recovers" the coordination by electrostatic interaction to relax. The 3-coordinated Fe1 of the (100) surface, mainly attracted by the S1, relaxes inward along the Fe–S bond. The 3-coordinated S1 is attracted by Fe1 and repelled by S in the second layer. Relaxing outward along the Fe–S bond causes the polyhedron to shrink, but without significant reconstitution.

Theoretically, depending on the breakage of the S on the surface, the number of coverage is different (0–8/single cell), and the (311) surface will have nine possible terminals. To get closer to the more realistic pyrite surface, the (311) supercell model with fractures in both the Fe–S and the S–S bonds is established (see Fig. 7.22).

The supercell model of the (311) surface is hexagonal with the size of $a_0 = 13.266436$, $b_0 = 15.318761$, and $c_0 = 19.674459$. Fe in the outermost layer is 3-coordinated. The 4-coordinated Fe and 5-coordinated Fe alternate horizontally in the atoms of the second layer, forming triangular pyramids and quadrangular pyramids on the surface, respectively. The 3-coordinated S in the outermost layer has two forms. One is the Fe–S bond breakage, in which the S is connected with two Fe and one S (see S1 in Fig. 7.22A). The other is the S–S bond breakage, where the Fe–S bond is not broken, with the S is connected to three Fe (see S2 in Fig. 7.22A).

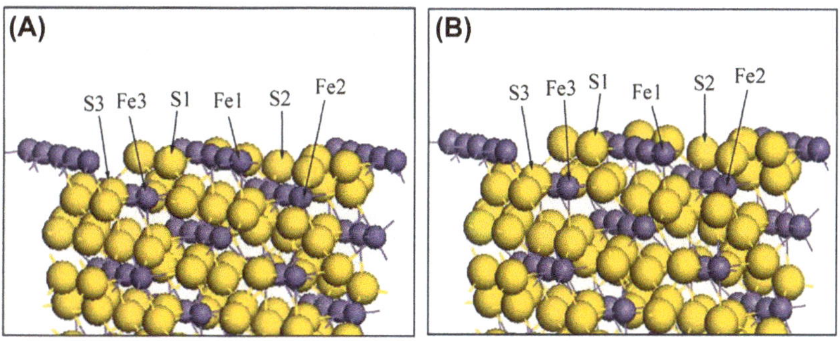

FIGURE 7.22 Pyrite FeS_2 (311) surface. (A) Before relaxation. (B) After relaxation.

TABLE 7.5 Pyrite (311) surface atomic coordination and displacements.

Atom label	Coordination number	Charge number (e)	Atomic displacement (Å)		
			Δ_x [0 0 1]	Δ_y [0 1 0]	Δ_z [1 0 0]
S1	3	−0.12	0.20	0.14	−0.05
S2	3	0.06	0.11	−0.06	−0.11
S3	4	0.08	0.03	0.04	−0.01
Fe1	3	0.14	−0.26	0.04	−0.12
Fe2	4	−0.03	−0.08	−0.10	−0.08
Fe3	5	−0.06	−0.01	−0.09	−0.06
Bulk Fe	6	−0.10	/	/	/
Bulk S	4	0.08	/	/	/

Table 7.5 shows that the 3-coordinated Fe (Fe1) in the outermost layer of the (311) surface has the most obvious relaxation, and relaxes inward along the Fe—S bond. The 4-coordinated Fe and 5-coordinated Fe relax along the Fe—S bond axis to the polyhedral vertex. The relaxation of the S1 is also obvious, making the S—S bond tilt parallel to the (311) surface. The relaxation of the S2 is relatively slight. For the Fe of the (311) surface, the lower coordination means the greater relaxation. Although S is 3-coordinated, S2 lacks the repulsion of the connected S due to the S—S bond breakage. In addition, its relaxation is weaker than that of the S1. There is no obvious reconstitution on the (311) surface.

The fresh surface formed by the pyrite after cutting, grinding, and polishing was characterized by AFM at the atomic level (see Fig. 7.23).

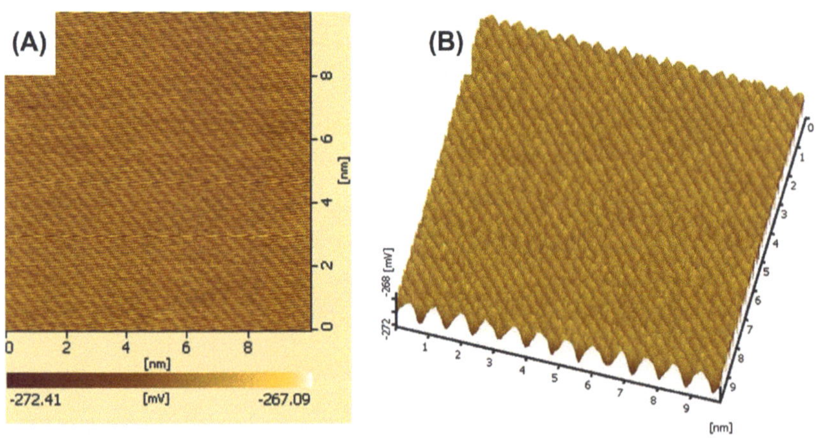

FIGURE 7.23 Atomic force microscopy topological image of pyrite surface. (A) Plan image. (B) Three-dimensional stereo image.

Fig. 7.23A shows the AFM plan of the pyrite surface. The atomic arrangement of the pyrite surface is obviously striped. The higher brightness stripe is the atomic arrangement in the outermost layer; the lower brightness stripe is the atomic arrangement below the surface layer; the atoms are arranged in parallel along the horizontal 135 degrees. The diameter of the atom is about 0.3 nm, which is slightly higher than the scale for S. It is close to the diameter of S after it obtains a negative charge. Therefore, the surface relaxation of pyrite causes the sulfur ions to lean outward.

Fig. 7.23B shows the AFM three-dimensional image of the pyrite surface. At the atomic scale, the pyrite surface is extremely uneven, and two identical atoms do not exist. It shows that the atomic arrangement of the pyrite surface has certain differences compared with the regular arrangement of atoms inside the crystal. In addition, when the surface atoms relax, the state of each atom changes, indicating that the atoms are not stationary. On the other hand, in the spatial normal direction, the heights of the surface atoms are inconsistent and can be considered as the range of displacement when the atoms are relaxed. The difference between the highest and the lowest is about 0.4 nm, which is larger than the calculated value by quantum chemistry. This may be due to the presence of foreign substances when the tests are conducted under nonvacuum conditions.

On the ideal pyrite surface, there is an uneven atomic order of magnitude, which indicates uneven arrangements of atoms on the pyrite surface in the spatial normal direction. That is, some surface atoms are displaced in the spatial normal direction relative to their original positions, namely through surface relaxation. When the pyrite is broken by an external force, a new surface is formed in a moment, and the chemical bond breakage is unstable due to high energy. To reduce the surface energy, the broken chemical bonds easily adsorb atoms of foreign substances. Even in a vacuum, the chemical bonds from broken surfaces reduce their energy by bond formation. This changes the positions of the surface atoms relative to their original regular position, thus causing relaxation. The relaxation causes the S-rich pyrite surface, which is consistent with the calculation results by quantum chemistry.

7.2.3 Interaction of chalcopyrite surface and the copper in inclusions

Copper ion is an important component of fluid inclusions in chalcopyrite and its associated minerals. Copper salts (such as copper sulfate and copper chloride) in actual mineral flotation are often used as flotation activators. To avoid the influence of the charging system on the energy calculation, copper chloride is used to improve the calculation accuracy in this section. Fig. 7.24 shows the most stable structure of $CuCl_2$ adsorption on a chalcopyrite (0 0 1) surface. Regardless of the initial position of $CuCl_2$, Cu interacts with the S on the chalcopyrite surface after adsorption, with the bond length ranging from

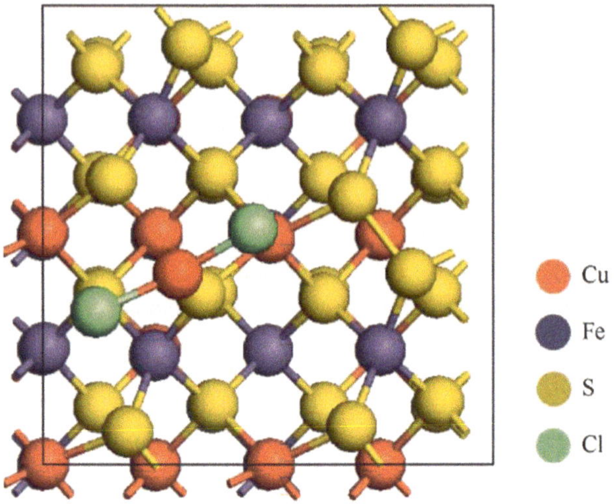

FIGURE 7.24 $CuCl_2$ stable adsorption structure on the chalcopyrite (0 0 1) surface.

2.221 to 2.405 Å, and the adsorption energy ranging from −23.986 to −20.966 kJ/mol. Neither of the chlorine atoms is bonded with the surface atoms.

Fig. 7.25 shows the structural parameters. The adsorption energy of the model is −23.986 kJ/mol. The Cl−Cu−Cl bond angle is 180 degrees before adsorption, and becomes 154.720 degrees after adsorption. The Cu−Cl bond length before adsorption is 2.053 Å, and becomes 2.119 Å after adsorption. The

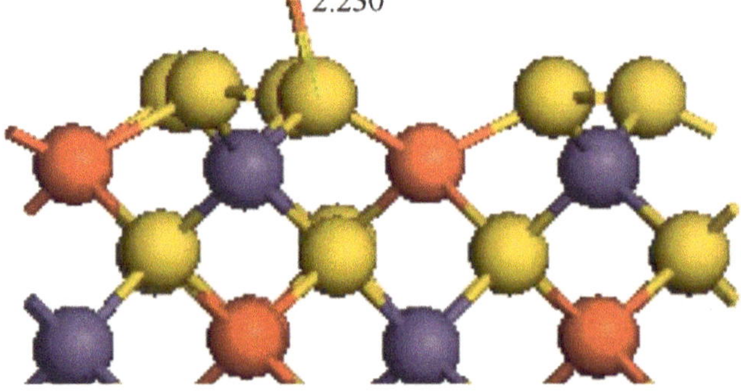

FIGURE 7.25 Adsorption structural parameters of the $CuCl_2$ and chalcopyrite (0 0 1) surface.

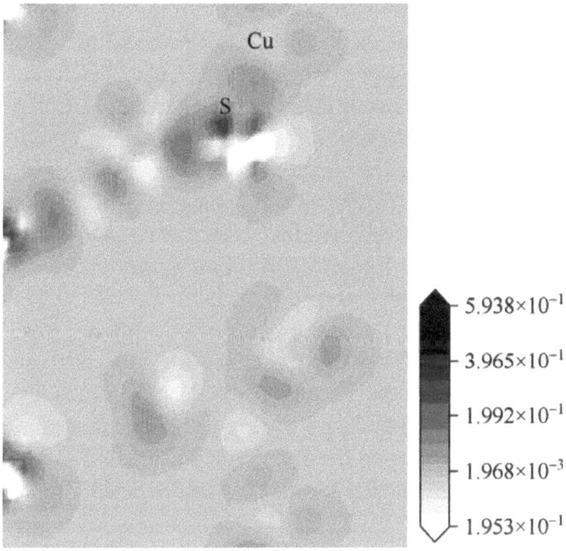

FIGURE 7.26 Difference charge density of $CuCl_2$ adsorbed on a chalcopyrite (001) surface.

Cu—S (surface) bond length is 2.230 Å, close to the theoretical Cu—S bond length of 2.300 Å. Combined with the adsorption energy, the copper chloride interacts with the surface. In addition, the adsorption further causes the surface of the S layer to rise, which is favorable for hydrophobicity and floatability.

Based on the surface properties and the electronic structure results of the adsorption system, the adsorption and bonding properties of $CuCl_2$ on the chalcopyrite (0 0 1) surface were further studied. The Mulliken charge population shows that the Mulliken charge of Cu in $CuCl_2$ before and after the adsorption is 0.26 and 0.40 e, respectively. This indicates that there is electron exchange between Cu and the surface S. During the adsorption, electrons can be transferred from the surface S to $CuCl_2$.

Fig. 7.26 shows the differential charge density of $CuCl_2$ and the chalcopyrite surface adsorption system. The interface in the figure passes through the Cu in $CuCl_2$ and the surface S, perpendicular to the surface. The local distribution of the valence electrons can be visually observed—the light color represents a region with rare electron density, and the dark color represents a region with high electron density. More electrons are distributed around Cu and S, which is consistent with the results of previous structural analysis.

7.2.4 Interaction of sphalerite surface and the copper in inclusions

Cu—Zn sulfide minerals are from skarn-type, hydrothermal-type, and hydrothermal-filling-and-metasomatic-type deposits, and have a large number of fluid inclusions. Due to the close co-consistent relationship between the chemical diversity of the ore-forming fluids and minerals in the ore formation, some ore-forming fluids are captured in or between host crystals or in or between host crystal and gangue mineral. The Cu in the fluid inclusions of sulfide mineral and its gangue mineral interacts with the sphalerite surface, which in turn activates flotation of sphalerite during the separation of Cu–Zn mineral. There are two forms of interaction between the Cu and the sphalerite. One is the replacement of the Zn on the sphalerite surface, and the other is adsorption on the S site of the sphalerite surface. The latter is an intermediate form of the former, and the former substitution is the final form of the interaction between Cu and the sphalerite surface. In this section, they are simulated.

Fig. 7.27 shows that the Zn in the two positions of the sphalerite surface can be substituted by Cu. They are Zn at the top (T) and bottom (B) sites. After the Cu substitutes for the Zn on the sphalerite surface, it forms a stable coordination with the surrounding S. The Cu–S bond formed by the Cu substitution for the T-site Zn on the sphalerite surface is 3-coordinated, while the Cu–S bond formed by the Cu substitution of the B-site Zn on the surface is 4-coordinated.

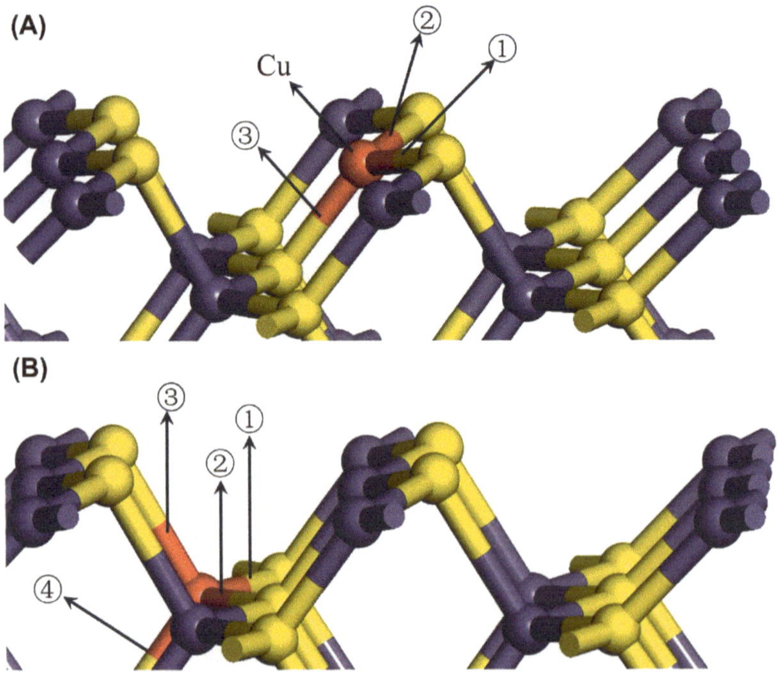

FIGURE 7.27 Configuration of Cu substitution for various Zn on the ZnS (110) surface. Cu substitution for (A) top-site Zn and (B) bottom-site Zn. ①–④ correspond to the formed bonds of Cu–S.

TABLE 7.6 Substitution energy, bond length of Cu−S, and bandgap after Cu substitution for Zn.

Site of Cu substitution	ΔE_{sub} (kJ·mol^{-1})	D_{Cu-S} (Å)	Bandgap (eV)
Top-site Zn	−47.40	①2.21, ②2.23, ③2.22	0.07
Bottom-site Zn	−31.34	①2.32, ②2.29, ③2.26, ④2.37	0.01

$D_{Cu\text{-}S}$, Cu−S bond length.

Table 7.6 shows that the substitution energy of Cu for Zn at the different sites on the sphalerite surface is negative, which indicates that the process is spontaneous. The quantum chemistry proves that there is indeed an ion exchange process during the Cu activation. The different substitution energy indicates the activation complex for substituting Zn at the different sites. The Cu substitution energy for T-site Zn on the sphalerite surface is the lowest (−47.40 kJ mol^{-1}), indicating that Cu is more likely to substitute the T-site Zn. Compared with the length of the Cu−S bond formed after the Cu substitution, the lengths of the Cu−S bond formed after the Cu substitution are slightly different from each other. Both are smaller than the ionic radius (2.57 Å) between the Cu and the S, but close to the covalent radius (2.19 Å) between Cu and S, indicating that the Cu−S covalent bond has polarity.

The DFT calculation and the AFM experimental results for the atomic relaxation on the sphalerite surface indicate that the atomic relaxation spontaneously occurs perpendicular to the normal direction of the sphalerite surface. The surface S is displaced outward, and Zn displaced inward into the crystal to form a relatively S-rich surface. In addition, the atomic radius of S^{2-} (184 pm) is much larger than that of Zn^{2+} (74 pm), so S is located in the outermost layer of the surface. The unique surface atomic structure of the sphalerite enables the surface S to adsorb copper ions. Fig. 7.28 shows the possible adsorption configuration of the copper ions on the sphalerite (110) surface after structural optimization.

As shown in Fig. 7.28, the copper ions may have five adsorption configurations for the S in the first layer on the sphalerite (110) surface:

The DFT calculation results show that Cu is significantly repelled by S when Cu and S interact in the Cu adsorption configuration (see Fig. 7.28C−E). After the structural optimization and energy convergence, the Cu spontaneously approaches the T-site S. The distance between the Cu and the specified S is larger than that between the Cu and the surface T-site S, and the length of the Cu−S bond formed by Cu and the specified S is larger than the ionic radius (2.57 Å) between Cu and S. These results indicate that the copper ions are more likely to interact with the T-site S on the sphalerite surface, but do not exclude the physical adsorption of Cu on S at other sites.

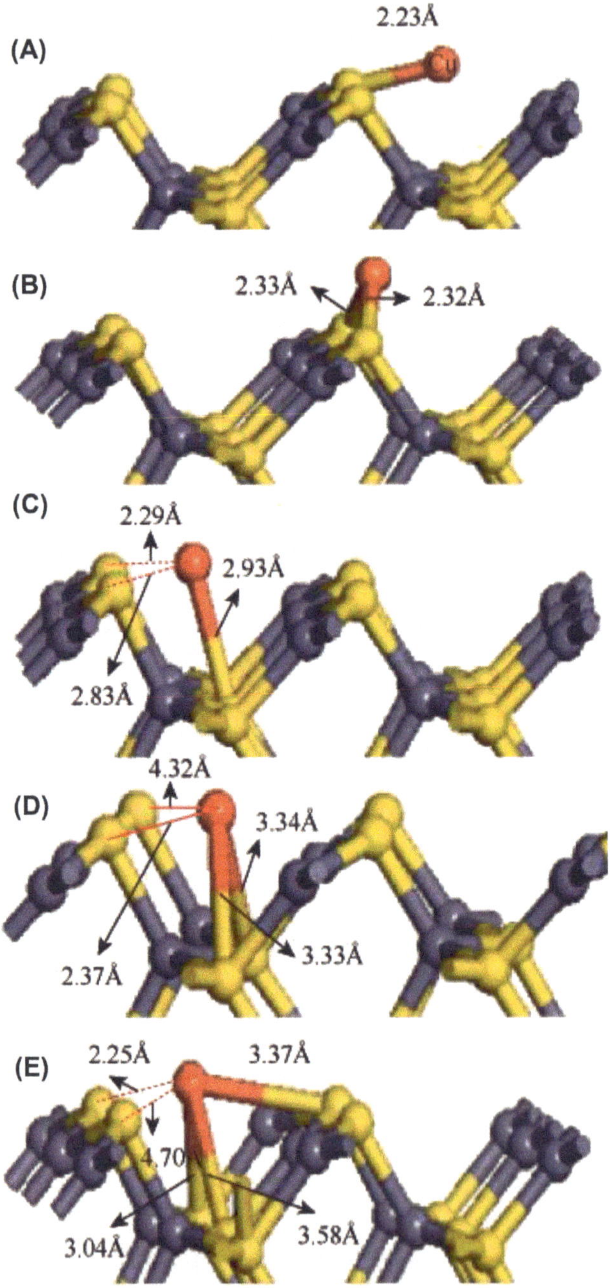

FIGURE 7.28 Configuration of Cu adsorption on the ZnS (110) surface. Cu adsorption on the (A) top (T) site of T-site S, (B) bridge site of T-site S, (C) top site of bottom (B)-site S, (D) bridge site of B-site S, and (E) threefold-hollow site of T- and B-site S.

Taking the Cu bridge adsorption on the B-site S (Fig. 7.28D) as an example, the interaction distances among Cu and two B-site S after the adsorption equilibrium are 3.33 and 3.34, respectively, which are much larger than the ionic radius (2.57 Å) between Cu and S. However, the distance between Cu and the T-site S near the outer side is 2.37 Å, indicating that the interaction between Cu and the T-site S is stronger than that with B-site S. Therefore, we discuss only the Cu adsorption on the T-site S of the sphalerite (110) surface.

Fig. 7.28A and B shows that the Cu−S bond formed by the Cu top-site adsorption on the T-site S is 2.23 Å, and the lengths of the two Cu−S bonds formed by the bridge-site adsorption are each 2.33 Å. Although there are some differences in the Cu−S bond lengths in the two adsorption configurations, they are all between the Cu−S covalent radius (2.19 Å) and the ionic radius (2.57 Å). This indicates that the copper ions have a strong chemical adsorption on the T-site S of sphalerite. Sigang Luo [1] used molecular mechanics to study the activation mechanism of copper ions on the sphalerite surface, finding that there is a bonding effect between the copper ions and the S on the sphalerite surface.

Table 7.7 shows the adsorption energy of Cu in the different adsorption configurations on the T-site S of the sphalerite (110) surface (see Fig. 7.28A and B) as well as the sphalerite bandgap after adsorption.

Table 7.7 shows that the ΔE_{ads} of the top- and bridge-site adsorptions of Cu on the T-site S of the sphalerite surface is 657.24 and −670.11 kJ/mol, respectively, further indicating that the adsorption of copper ion on S is a chemical process. The negative adsorption energy indicates that the adsorption is a spontaneous reaction. The ΔE_{ads} of the bridge-site adsorption is significantly larger than that of top-site adsorption, which indicates that the bridge-site adsorption of Cu on the two S is more stable. In addition, the bandgap of the sphalerite after Cu adsorption is significantly reduced from the original 2.71 to 0.07 (top-site adsorption) and 0.09 eV (bridge-site adsorption), indicating that the conductivity of the sphalerite surface increases significantly after Cu adsorption.

Fig. 7.29 shows that the density of state of the atoms on the sphalerite surface changes after Cu adsorption. Both adsorptions lead to the formation of Cu 3*d* orbital peaks, which are located near −1 eV of Fermi level. This indicates that the Cu possesses a strong activity, which is beneficial to the

TABLE 7.7 Adsorption energy and bandgap after Cu adsorption on the S site of the ZnS surface.

Cu adsorption site	Cu adsorption type	ΔE_{ads}/kJ/mol	Bandgap (eV)
T-site S	Top	−657.24	0.07
	Bridge	−670.11	0.09

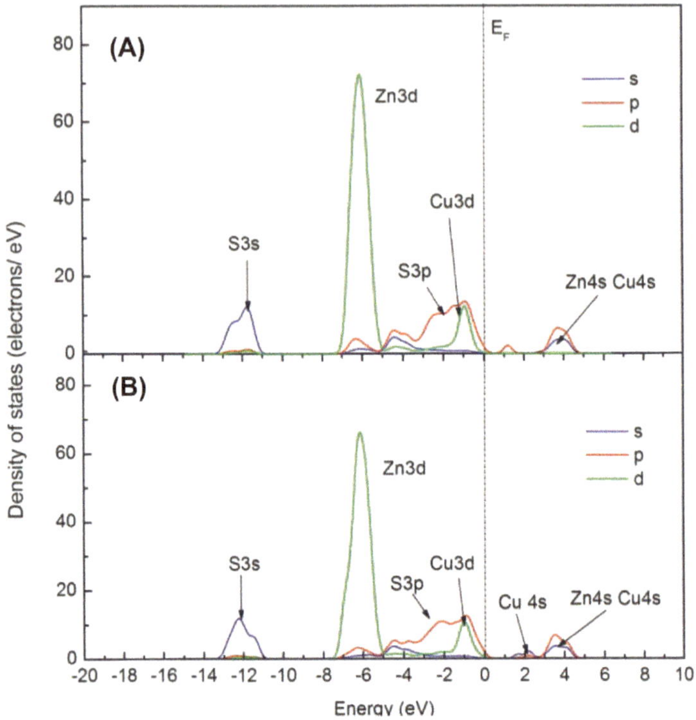

FIGURE 7.29 Density of state of the surface atoms of the ZnS (110) surface after Cu adsorption. (A) Top- and (B) bridge-site adsorption.

collectors. For both adsorptions, the Cu 3*d* orbit and the S 3*p* orbit overlap between −2.00 and 0 eV. The hybridization between the orbits shows that the adsorption between Cu and S should be a chemical adsorption. In addition, the 4*s* orbital electrons of Cu adsorption contribute to the conduction band.

Table 7.8 shows that both the top-site adsorption of Cu on the T-site S and the bridge-site adsorption between the two T-site S result in the transfer of electrons from S to Cu. That is, after Cu adsorption, the Cu on the sphalerite surface gains electrons, while the S loses electrons. When the Cu accepts electrons, the valence decreases; when the S loses electrons, the valence increases and oxidizes. Taking the top-site adsorption of Cu on the sphalerite surface as an example, the charge of Cu decreases from +2 to +0.36 e after adsorption, while the charge of S increases from −0.48 to −0.44 e.

During Cu adsorption, the *s*, *p*, and *d* orbits of Cu mainly derive electrons from the *p* orbit of the S. Zn loses some electrons during the adsorption, which may a result of the calculation error of the system. In the actual system, Cu and Zn should not interact. In summary, there is electron transfer between Cu and S during the adsorption of Cu on the sphalerite surface. It is a process in which Cu is reduced, and S is correspondingly oxidized. Fig. 7.30 shows the optimal

TABLE 7.8 Average Mulliken population of atoms before and after Cu adsorption on S site.

Interaction type		Atomic type	Population number of atomic orbit			Total	Charge (e)
			s	*p*	*d*		
Before Cu adsorption	—						
		S	1.83	4.66	0.00	6.49	−0.48
		Zn	0.69	0.85	9.98	11.52	0.48
		Cu	0.00	0.00	9.00	9.00	2.00
After Cu adsorption	Top	S	1.83	4.61	0.00	6.44	−0.44
		Zn	0.68	0.84	9.97	11.49	0.51
		Cu	0.56	0.20	9.88	10.64	0.36
	Bridge	S	1.83	4.62	0.00	6.45	−0.45
		Zn	0.68	0.83	9.97	11.48	0.52
		Cu	0.50	0.24	9.86	10.61	0.39

The average population of S after Cu adsorption refers to the population mean of the S coordinated with Cu.

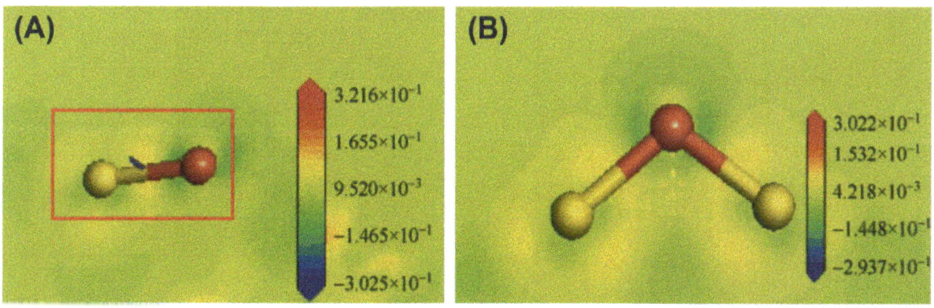

FIGURE 7.30 Difference in electron densities of ZnS (110) surface after Cu adsorption. (A) Top- and (B) bridge-site adsorption.

field of view of the differential charge density formed by segmenting along the Cu−S bond after the Cu adsorption on the T-site S of the sphalerite (110) surface.

Fig. 7.30 shows that along the Cu−S bond, there is a small overlap between the electron clouds of the two atoms, which further indicates the interaction. The charge density around S is significantly stronger than that of Cu.

7.2.5 Interaction of the pyrite surface and the Cu in inclusions

First, we studied the interaction of Cu on the pyrite surface—the copper ion substitution for the top Fe-site Fe1 and the two adsorptions of copper ions on the top S-site S1 (Fig. 7.31). Before the interaction with S1, the copper ions were placed in the normal direction 4 Å away from S1.

The results showed that after the Cu substitution of the Fe, a 3-coordinated polymer formed. The coordinated S was biased to one side, with poor structural balance. The substitution energy (238.23 kJ/mol) of the Cu for Fe was positive, indicating that Cu did not substitute for Fe. Therefore, Cu adsorption on the pyrite surface was not achieved by substituting Fe. The adsorption energy of the copper ions directly adsorbed on the surface S1 was −757.85 kJ/mol, which tended to be chemical adsorption. Moreover, the Cu and S bonded, with stable adsorption. Theoretical calculations showed that Cu adsorption on the pyrite surface is a direct S-site adsorption rather than substitution.

Fig. 7.32 shows the charge density after copper ion is adsorbed on S1 of the pyrite surface. The electron-overlapping density between the copper ion and S1 is large, and the Cu−S bond length is 2.230 Å, with strong covalency.

Fig. 7.33 shows the charge density difference after copper ion adsorption on S1 of the pyrite (100) surface. The copper ions are adsorbed with electrons

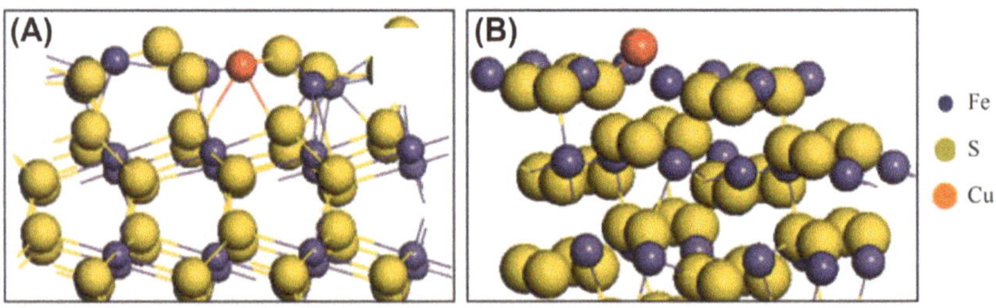

FIGURE 7.31 Two adsorptions of Cu on the pyrite (100) surface. (A) Cu substitution for Fe1 and (B) Cu adsorption on surface S1.

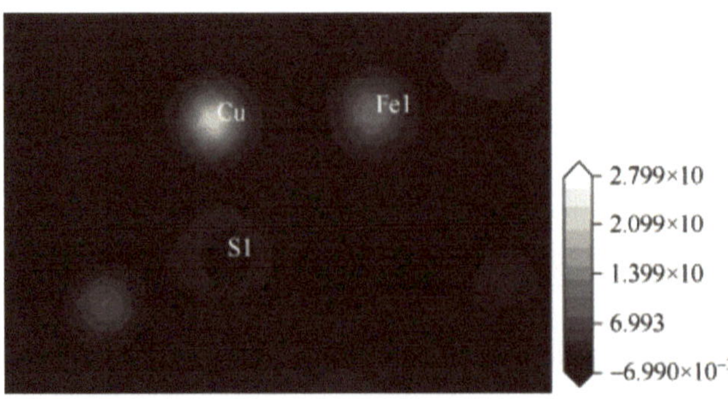

FIGURE 7.32 Electron densities of atoms on the pyrite (100) surface after copper ion adsorption.

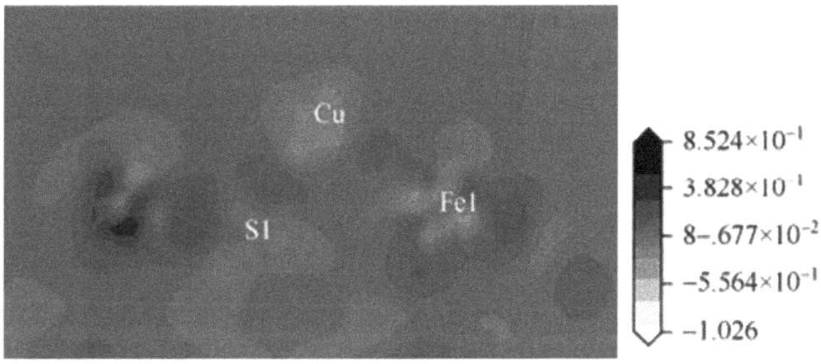

FIGURE 7.33 Difference in electron densities of atoms on pyrite (100) surface after copper ion adsorption.

TABLE 7.9 Mulliken charge populations of atoms before and after copper ion adsorption.

		Orbital population				
Atom label	Status	*s*	*p*	*d*	Total	Charge (e)
Fe1	Before adsorption	0.39	0.36	7.09	7.86	0.14
	After adsorption	0.39	0.43	7.09	7.91	0.09
S1	Before adsorption	1.81	4.33	0	6.15	−0.15
	After adsorption	1.79	4.24	0	6.07	−0.03
Cu	Before adsorption	0	0	0	9.00	2.00
	After adsorption	0.95	0.34	9.74	11.03	−0.03

accumulated, and the electron cloud around S1 is missing, indicating that electrons are transferred between copper ions and S1. The electrons should be transferred from S1 to the copper ions. Table 7.9 shows the population values of the atomic charge before and after the adsorption of copper ions on the pyrite (100) surface. The atomic charge population also shows the charge transfer between the atoms. After the adsorption of copper ions, the 3*s* orbit of the surface S1 loses some electrons, while the 3*p* orbit loses more. The charge number of S1 changes from −0.15 to −0.03 e before the adsorption. Fe1 obtains electrons mainly from the 3*d* orbit, and the charge number is changed from 0.14 to 0.16 e before the adsorption. The calculation results show that copper ions have weak substitution for the surface Fe.

7.3 Interactions among sulfide mineral surface, components of fluid inclusions, and collectors

The newly formed surface of the sulfide mineral spontaneously undergoes relaxation and reconstruction of the surface atoms. As a result, the surface S moves outward, and the metal atoms move into the crystal to form a relatively S-rich surface. In addition, the radius of the S is much larger than those of the metal atoms such as Cu, Pb, and Zn. The unique surface structure of the sulfide mineral gives it a natural hydrophobicity. However, the sulfide mineral surface has a natural adsorption tendency for the metal ions in the ore pulp solution.

During crushing and grinding of the ore, the decrease in the mineral particles leads to a large number of newly formed mineral surfaces, as well as the release of fluid inclusion components. The released inclusion components, especially the active components such as Cu, are strongly adsorbed on the newly formed mineral surfaces with high activity. The preactivation on the surface of the sulfide mineral in turn affects the selectivity of the collectors, which makes the flotation separation of polymetallic sulfide mineral difficult. The interactions among chalcopyrite, Cu, and collectors, as well as those among sphalerite, Cu, and the collectors, are taken as examples to illustrate the possible interaction models and their microscopic mechanisms.

7.3.1 Interaction of chalcopyrite surface, Cu components, and xanthate

After the copper ions of chalcopyrite fluids are released into the solution, the possible physical and chemical effects include a series of balance reactions of copper ions in the aqueous solution, as well as the adsorption and desorption of copper ions on the chalcopyrite surface. During the flotation process, the xanthate dissociates in the water to form xanthonic acid ions, which involves the interaction of copper ions, xanthonic acid, and the chalcopyrite surface.

The process happens in two stages: One is the chemical reaction in solution, that is, copper ions react with xanthate to form Cu−xanthate, which will not be discussed here. The other refers to adsorption or chemical reaction on the mineral surface. The copper ions react with the chalcopyrite surface to form Cu−S, and then the xanthonic acid ions are adsorbed on the Cu. Both forms promote the flotation of chalcopyrite under certain conditions. Therefore, this process occurs during grinding and dissociation of the minerals, and the release of the inclusion components is mentioned as self-activation. The activation behavior induced by the surface adsorption on the components of fluid inclusions in chalcopyrite in this book is called self-activation.

The chalcopyrite surface−Cu−xanthate adsorption model used is an optimized $2 \times 2 \times 2$ supercell structure of chalcopyrite, taking the surface-terminated atomic layer as the metal layer. The possible form of the ternary structure of the charged system was first tested by Gaussian. The results showed the stereo structure of the xanthogen−Cu−cluster model, and then the VASP adsorption model was established (see Fig. 7.34).

The results showed that a bonding existed between Cu and two S on the surface. The Cu was located above the middle of the two S and interacted with more than one S. The copper ion was located between the S in the ethyl

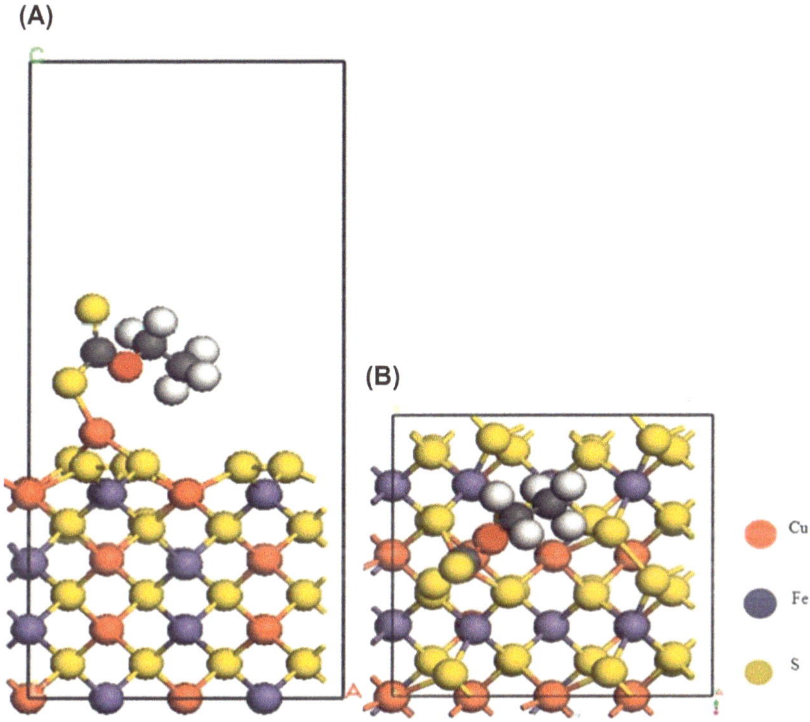

FIGURE 7.34 Chalcopyrite (0 0 1) surface S—Cu—ethyl xanthate ternary structure calculation models. (A) Side view, (B) view perpendicular to the surface.

xanthate (EX) and the S on the chalcopyrite surface, with a bridging effect. The bond lengths between the Cu and the two surface S were 2.186 and 2.209 Å, respectively. However, compared with the case of no xanthate, the bond length of the Cu and the surface S was short due to the electron exchange between the S—Cu in EX and the ternary system of the surface S. The other S in the EX molecule had a longer distance from Cu, with little effect on the surface. In addition, after adsorption, the chalcopyrite (0 0 1) surface produced a partial relaxation, and the S—S and S—Cu bonds on the surface were different from those in the sublevel. The adsorption energy of this configuration was −804.687 kJ/mol, similar to that of Cu-free EX on each adsorption potential of the chalcopyrite surface. However, it was larger than the adsorption energy of copper ions on the chalcopyrite surface, indicating that the copper ions had an important influence on the adsorption of xanthate and the chalcopyrite surface.

Fig. 7.35 shows the charge density difference of the S—Cu—EX ternary structure formed by the adsorption of collectors on the chalcopyrite surface. The interface passes through EX and the surface S, perpendicular to the surface. There is overlap of the electron clouds among S in EX, Cu, and the surface atoms, which shows the interaction of the ternary system.

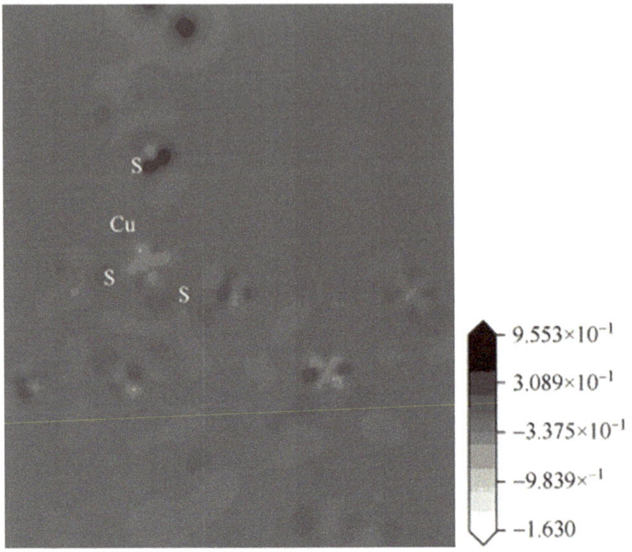

FIGURE 7.35 Charge density difference map of ethyl xanthate—Cu—S.

Copper ion, one of the main components in the fluid inclusions of chalcopyrite, is adsorbed on chalcopyrite, mainly by the adsorption of copper ions on surface S. This has been proven in previous experiments and theoretical studies. That is to say, the copper ions of fluid inclusions in chalcopyrite have activation effects on sphalerite. Based on experimental and theoretical calculations, the components released from fluid inclusions are an important factor. Fig. 7.36 shows the induced activation flotation paths of copper ions released from fluid inclusions. On one hand, the copper ions react with the

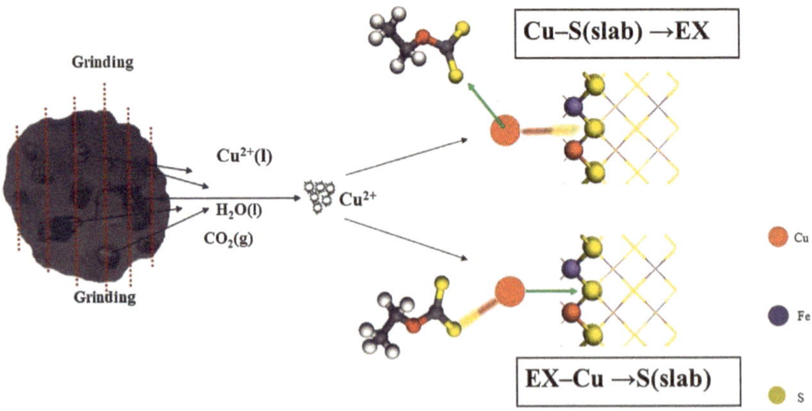

FIGURE 7.36 Induced activation flotation paths of copper ions released from fluid inclusion. *EX*, ethyl xanthate.

collector ions to form copper xanthate, which is adsorbed on the surface S to form a hydrophobic surface. On the other hand, the copper ions first interact with the surface S to form Cu_xS_y, and the collector ions interact with these Cu-based sulfides to form a hydrophobic surface.

7.3.2 Interaction of sphalerite surface, Cu, and xanthate

As described in the previous chapters, there is a close interassociated relationship between minerals in the ore-forming process and the chemical diversity of the ore-forming fluids. The mutual ore-forming fluids exist in chalcopyrite, sphalerite, and pyrite produced from the deposits of polymetallic sulfide minerals as well as the associated fluid inclusions of the gangue minerals. During the grinding of polymetallic sulfide mineral, the Cu released by the fluid inclusions inevitably preactivates sphalerite, which is detrimental to the separation of sphalerite from other metal sulfides. The interactions among the sphalerite−Cu−EX ternary systems were calculated, and Fig. 7.37 shows the model after the structural optimization. After the Cu released by the fluid inclusion interacts with the sphalerite surface, the EX ion is bonded to the surface Cu of the sphalerite in a vertical manner, and the hydrophobic group is located on the outermost surface.

Theoretical calculations show that there are four stable interaction modes among sphalerite, Cu, and EX: (1) EX and sphalerite surface with substituted Cu, (2) EX and Cu adsorbed on the top site of S, (3) EX and Cu adsorbed on the bridge site of S, and (4) EX and $Cu(OH)_2$ adsorbed on the surface. These can activate the flotation of sphalerite. In particular, when EX interacts with the $Cu(OH)_2$ adsorbed on the surface (see Fig. 7.37D), OH^- in the $Cu(OH)_2$ interacts with the Zn at some sites of the sphalerite surface. Cu−OH bond breakage in the molecule releases free OH^- (see the dotted circle in the figure), and the Cu in the molecule interacts with the sphalerite surface. Therefore, the essence of the interaction between EX and surface $Cu(OH)_2$ is that EX interacts with the Cu in the dissociated or partially dissociated $Cu(OH)_2$, which interacts with the sphalerite surface S to activate flotation. Cu interacts as bridges throughout the activated flotation.

The DFT results show that after the interaction of the sphalerite−Cu−EX ternary medium, there is little change in the bond length between the atoms in EX, but the bond angle changes greatly (see Table 7.10).

Table 7.10 shows that when EX interacts with Cu on the sphalerite surface, its molecular structure changes. This structural change is mainly reflected in the changes of the bond angle, which are dissimilar under different interaction models. After the interaction, some bond angles reduce, while some enlarge. However, the common point is that the bond angle ($\angle$S1=C3−S2), formed by S, double bond S, and C, which are bonded with Cu, is enlarged after the interaction. Table 7.11 shows the interaction energy (ΔE_{int}) of the four models

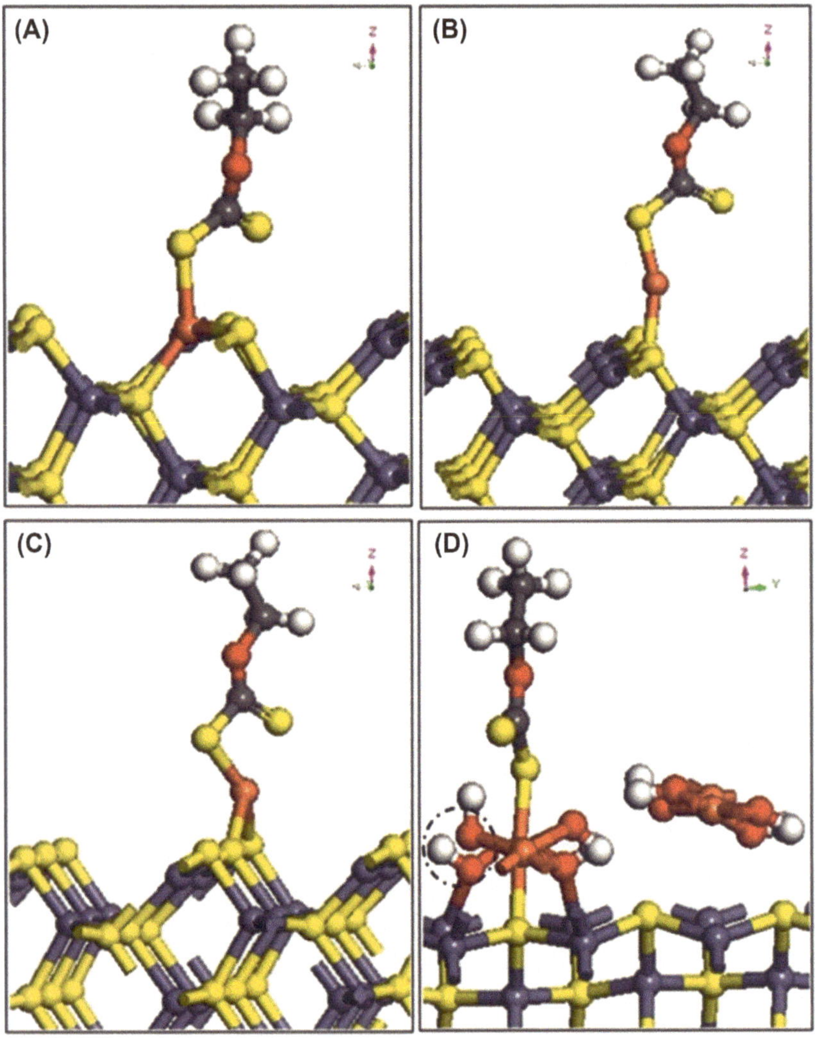

FIGURE 7.37 Ternary media interaction model of sphalerite ZnS (110) surface—Cu—EX. EX with (A) substituted Cu, (B) Cu adsorbed on the top site of S, (C) Cu adsorbed on the bridge site of S, and (D) $Cu(OH)_2$ adsorbed on the ZnS surface. *EX*, ethyl xanthate.

of the sphalerite—Cu—EX ternary medium, as well as the Cu—S bond length (D_{Cu-S}) formed between the bonded S in EX and the Cu on the sphalerite surface. ΔE_{int} is calculated by Eq. (7.7).

Table 7.11 shows that the ΔE_{int} of each model is much larger than zero, indicating that the four models interact spontaneously. At the same time, the ΔE_{int} of the four Cu-activated flotation models are much larger than that (−87.63 kJ/mol) of the direct interaction of EX and the Zn of the sphalerite

TABLE 7.10 Bond angle of ethyl xanthate before and after interaction with Cu-activated ZnS (110) surface.

Type of Cu interaction	EX	Bond angle (degrees) ∠C1–C2–O	∠C2–O–C3	∠O–C3–S2	∠O–C3=S1	∠S1=C3–S2
Substitution	Before action	108.46	116.06	126.66	122.07	111.76
	After action	105.51	123.52	110.26	119.89	129.85
	Increased or decreased value	−2.95	+7.46	−16.40	−2.18	+18.09
Top-site adsorption on S	After interaction	106.46	121.88	112.53	126.92	120.36
	Increased or decreased value	−2.00	+5.82	−14.13	+4.85	+8.60
Bridge-site adsorption on S	After interaction	106.44	125.15	117.19	128.04	114.77
	Increased or decreased value	−2.02	+9.09	−9.47	+5.97	+3.01
$Cu(OH)_2$ adsorption	After interaction	106.35	120.49	108.55	123.27	128.17
	Increased or decreased value	−2.11	+4.43	−18.11	+1.2	+16.41

EX, ethyl xanthate.

TABLE 7.11 Adsorption energy and bond length of Cu−S for the 4 interaction models.

Interaction model	ΔE_{int} (kJ/mol)	D_{Cu-S} (Å)
ZnS(110)−substituted Cu−EX	−232.52	2.24
ZnS(110)−Cu adsorbed on top site of S−EX	−256.11	2.14
ZnS(110)−Cu adsorbed on bridge site of S−EX	−268.65	2.28
ZnS(110)−surface $Cu(OH)_2$−EX	−281. 60	2.31

$D_{Cu\text{-}S}$, Cu−S bond length; *EX*, ethyl xanthate.

surface. This indicates that after the Cu activation in sphalerite, the interaction between EX and surface Cu is easier than that between EX and Zn on the sphalerite surface.

In the four Cu-activated flotation models, the D_{Cu-S} formed between the S in EX and the Cu on the sphalerite surface is smaller than the ionic radius (2.57 Å) between Cu and S, and is close to/less than the covalent radius (2.19 Å) between Cu and S. This further indicates that the bonded S in EX chemically interacts with Cu on the sphalerite surface, and the Cu−S bond formed is a mixed bond of covalent and ionic bonds. Combining the results of the interaction of the sphalerite surface and the Cu in inclusions, we discussed the four kinds of Cu-activated flotation models and their applicable conditions for sphalerite, Cu in the fluid inclusions, and the collectors.

(1) Cu is substituted by EX and the sphalerite surface (see Fig. 7.36), suitable for the entire pH range. It becomes more obvious with longer activation time, and dominant with strong acidity in the solution.
(2) EX and Cu adsorbed on the S of the sphalerite surface (see Fig. 7.36) is suitable for weakly acidic and alkaline ore pulp solutions.
(3) EX and $Cu(OH)_2$ adsorbed on the surface directly interact in the flotation of the sphalerite (see Fig. 7.36), which is suitable for low Cu concentrations in the solution and short Cu activation time under alkaline conditions.

7.4 Interaction of the components released from mineral fluid inclusions in sulfide mineral flotation

In nature, Cu−Pb−Zn polymetallic sulfide minerals are closely associated with one another. According to the fluid chemistry of the mineral formation, the components of fluid inclusions in metal sulfide mineral should contain one of the homonymic heavy metal ions of the main components. For example, fluid inclusions of Cu sulfide ore, sphalerite, and galena contain Cu, Zn, and Pb, respectively.

Due to the close co-consistent relationship between minerals during ore formation and the chemical diversity of the ore-forming fluids, some ore-forming fluids are captured in or between the host crystals, or in or between the host crystal and the gangue mineral. In other words, the fluid inclusions in the Cu sulfide minerals produced in the polymetallic sulfide ore deposits may contain sphalerite; those of sphalerite may contain Cu sulfide (chalcopyrite); the gangue mineral may contain components of copper sulfide or sphalerite, or multiple chemical components of Cu, Pb, and Zn.

Fluid inclusions in minerals are small in size but contain a large number of components. In addition, the actual polymetallic sulfide mineral is in the majority of gangue minerals. If considering the influence of the inclusions, we should focus on the influence of the components released from the mineral inclusions on the chemical properties of the ore pulp solution and the mineral surface. The released components bring activation or inhibition in mineral flotation, which increases the complexity of the flotation solution systems and the questions regarding the flotation system, the interface chemistry, and the solution chemistry of flotation. We discuss the effects of the components released from mineral fluid inclusions on the flotation of sulfide mineral from the following three aspects.

First, we discuss the effects of the components released from the fluid inclusions on the flotation separation of sulfide mineral.

The flotation separation of Cu−Pb−Zn polymetallic sulfide mineral has been one of the most difficult and controversial topics in mineral processing around the world. The difficulty in separation is due to the complex mineral composition and mineral properties of polymetallic sulfide; one of the important reasons is the inevitable preactivation of inhibited minerals by heavy metal components in the ore pulp, especially the Cu−Pb activated component.

The introduction to this chapter indicates that the surface of the new sulfide mineral spontaneously undergoes relaxation and reconstitution of the surface atoms. As a result, the surface S move outward, and the metal atoms move inward into the crystal, to form a relatively S-rich surface. In addition, the radius of the S becomes larger than those of the metal atoms such as Cu, Pb, and Zn. The unique surface structure of the sulfide mineral gives it a certain natural hydrophobicity. However, the sulfide mineral surface also has a natural adsorption tendency for the metal ions in ore pulp solution.

Under the natural flotation pH conditions, the heavy metal components released by the fluid inclusions of sulfide minerals play a dominant role in flotation separation. During grinding, the components of inclusions in metal sulfide mineral and gangue mineral, especially the heavy metal activation components such as Cu and Pb, are released into solution where they interact with the surfaces of the new minerals with high activity. This results in the convergent surface of Cu−Pb−Zn sulfide mineral, namely surface homogenization. The surface of single sulfide mineral turns into a Cu−Pb−Zn surface of sulfide mineral, which leads to the difficulty in flotation separation. This is why sulfide mineral in some polymetallic sulfide deposits is difficult to float. Based on this, the surface and interfacial reactions of Cu−Zn sulfide mineral flotation can be regulated by inhibiting the release of copper ions, which is helpful to improve the selective flotation of Cu−Zn sulfide mineral. These new findings are closely linked to the flotation of fluid inclusions in Cu−Pb−Zn sulfide mineral.

Second, we discuss the influence of the components released from the fluid inclusions on the chemical system of the sulfide mineral flotation solution.

In Chapter 6, the theoretical solubility and dissolution characteristics of sulfide mineral are systematically described. Sulfide mineral has a small solubility product constant, with extremely low solubility. For example, the solubility of Cu−Zn sulfide mineral in neutral pH is in the range of

10^{-9}–10^{-15} mol/L, since only limited components of heavy metals are released during dissolution. Therefore, the components released from the fluid inclusions in metal sulfide mineral and its gangue minerals make an important contribution to the ore pulp solution in flotation; especially, in the neutral ore pulp system, the components that are released from the inclusions predominate. In addition to metal ions such as Na^+, K^+, Ca^{2+}, Mg^{2+}, Cu^{2+}, Pb^{2+}, and Zn^{2+}, mineral fluid inclusions are rich in anionic components (such as Cl^-, F^-, and SO_4^{2-}), halite, halides, sulfides, sulfates, carbonates, phosphates, silicates, borates, and metal oxides. As an important source of inevitable ions in flotation pulp, the inclusions release metal cations such as Cu^{2+}, Pb^{2+}, Zn^{2+}, Fe^{2+}, Ca^{2+}, Mg^{2+}, and K^+, which complicate the flotation system and destroy the selectivity of flotation.

The metal ions released by these inclusions undergo hydrolysis and complexation in the solution system, which affects or changes the separation. In addition to their adsorption on the mineral surface, metal ions interact with the collectors in the ore pulp solution. For example, Cu^{2+} and Pb^{2+} in the solution react with xanthate to form a precipitate of xanthated copper and consume the collector. Ca^{2+} and Mg^{2+} in the components of inclusions also affect the flotation of oxidized mineral. In addition, as inclusions are rich in chloride salts, which are likely to have an important impact on the flotation system, the actual flotation system may be carried out in aqueous solution containing salinity. The chloride in the flotation system has an influence on the flotation by affecting the dissolution rate of the mineral, electric double layer of the mineral–water interface, complexation of system metal ions, flotation foam properties, and even behaviors of the agents.

Third, we discuss the effects of components released from the fluid inclusions on the difference in mineral floatability.

Studies in geology show that the abundance and chemical composition of mineral fluid inclusions are significantly different in ore deposits from different regions due to the different metallogenic conditions. In production, it is often encountered that the same minerals produced by different mines or different mining areas in the same mine have different floatability. For some mines, it is easy to separate the Cu and Zn or Pb and Zn, while for some mines it is not.

In addition to the known disseminated grain size, mineral structure, and composition, the contents of fluid inclusions in minerals and their chemical composition are important factors. During the crushing and grinding of a mineral, the mineral grain size becomes smaller, and a large number of inclusions are opened and their fluid components are released, thus complicating the flotation system. The ions released by the inclusions in different deposits differ in type and quantity. These components have different activation or inhibition effects on the minerals, which may cause different flotation behaviors of the same minerals from different regions.

Reference

[1] Sigang Luo. Activation mechanism of sphalerite flotation in the presence of Cu^{2+} by molecular mechanics. Morden ming 2012;(3):7–9.

Further reading

[1] Kelebek S, Smith GW. Electrokinetic properties of a galena and chalcopyrite with the collectorless flotation behaviour. Colloids Surface 1989;40(0):137–43.

[2] Wang DZ, Hu YH. Solution chemistry of flotation. Changsha, China: Hunan Science & Technology Press; 1988.

[3] Jones MH, Woodcock JT, editors. Principles of mineral flotation, the Wark Symp, vol. 43, Victoria: Aust. Inst. Min. Metall; 1984.

[4] Tang LS, Huang KG, Wang DZ. Effect machanism of cooper ions on sulphie ores. Min Metall Eng 1989;3:29–32.

[5] Rath RK, Subramanian S. Adsorption, electrokinetic and differential flotation studies on sphalerite and galena using dextrin. Int J Miner Process 1999;57(4):265–83.

[6] Kresse G, Furthmüller J. Software VASP, Vienna (1999). Phys Rev B 1996;54(11):169.

[7] Kresse G, Furthmuller J. Vienna Ab-initio simulation package. Vienna: University of Wien; 2001. p. 1.

[8] Kresse G, Joubert D. From ultrasoft pseudopotentials to the projector augmented-wave method. Phys Rev B 1999;59(3):1758–75.

[9] Monkhorst HJ, Pack JD. Special points for Brillouin-zone integrations. Phys Rev B 1976;13(12):5188–92.

[10] Blöchl PE, Jepsen O, Andersen OK. Improved tetrahedron method for Brillouin-zone integrations. Phys Rev B 1994;49(23):16223.

[11] Feynman R. Forces in molecules. Phys Rev 1939;56(4):340.

[12] Perdew JP, Wang Y. Accurate and simple analytic representation of the electron-gas correlation energy. Phys Rev B 1992;45(23):13244.

[13] Martins JL, Troullier N, Wei SH. Pseudopotential plane-wave calculations for ZnS. Phys Rev B 1991;43(3):2213.

[14] Chen JH, Chen Y, Li YH. Effect of vacancy defects on electronic properties and activation of sphalerite (110) surface by first-principles. Trans Nonferrous Metals Soc China 2010;20(3):502–6.

[15] Chen JH, Chen Y. A first-principle study of the effect of vacancy defects and impurities on the adsorption of O_2 on sphalerite surfaces. Colloids Surface A 2010;363(1–3):56–63.

[16] de Lima GF, et al. Water adsorption on the reconstructed (001) chalcopyrite surfaces. J Phys Chem C 2011;115(21):10709–17.

[17] Klauber C. Fracture-induced reconstruction of a chalcopyrite ($CuFeS_2$) surface. Surf Interface Anal 2003;35(5):415–28.

[18] Harmer SL, et al. S species at chalcopyrite ($CuFeS_2$) fracture surfaces. Am Mineral 2004;89(7):1026–32.

[19] Luo W, Hu W, Xiao S. Size effect on the thermodynamic properties of silver nanoparticles. J Phys Chem C 2008;112(7):2359–69.

[20] Duke CB, Paton A, Kahn A. The atomic geometries of GaP (110) and ZnS (110) revisited: a structural ambiguity and its resolution. J Vac Sci Technol 1984;2(2):515–8.

[21] Dean JA. Lange's handbook of chemistry. Beijing: Science Press; 1985.

[22] Sun W, Hu YH, Qin WQ. DFT research on activation of sphalerite. Trans Nonferrous Metals Soc China 2004;14(2):376–82.

[23] von Oertzen,W GU, Skinner.H M, Nesbitt,A W, Pratt RA, Buckley N. Cu adsorption on pyrite (100): Ab initio and spectroscopic studies. Surf Sci 2007;601:5794–9.

Index

Note: 'Page numbers followed by "f" indicate figures, "t" indicates tables'.

A

Ammonium dichromate, 31
Atomic absorption spectroscopy, 28–29, 52–53

B

Bangong Lake-Nujiang metallogenic belt, 70–74
Bornite, 100
Broyden-Fletcher-Goldfarb-Shanno (BFGS) algorithm, 191

C

Calcite, 100, 100t, 114–118
Cambridge Serial Total Energy Package (CASTEP), 191
Carbon dioxide inclusions, 22
Car-Parrinello method, 190
Chalcopyrite, 220–223, 221f–222f
  chemical composition, 99, 99t
  crystal model and calculation method, 191–193, 192f
  crystal structure and purity, 99
  daughter mineral, 78
  high-resolution X-ray microfault three-dimensional imaging (HRXMT) analysis, 111, 111f–112f, 112t, 114t
  infrared optical microscopic analysis, 107–108, 107f
  quartz and calcite
    contact relations, 116, 116f
    Cu, Fe, Cl^-, SO^{2-}_4, 116–118, 117t
    and opaque minerals micrograph, 114–115, 115f
    optical microscopy imaging, 114, 115f
  scanning electron microscopy/energy-dispersive X-ray spectroscopy (SEM/EDS)
    inclusion domain, 109–110, 109f
    residual position domains, 108–109, 108f
    semiquantitative analysis, 109–110, 110t
  solubility
    acidic conditions, 155
    neutral and alkaline solutions, 155
    vs. pH, 153, 154f
    vs. time, 153, 154f
  sphalerite
    Cu, Fe, Cl^-, SO^{2-}_4, 121, 121t–122t
    flotation separation methods, 118–119
    fluid inclusion photomicrograph, 120, 120f
    fluid inclusion sheets, 119, 119f
    micrograph, 119, 119f–120f
  surface relaxation and reconstitution, 196f
    before and after optimization, 196–197, 197f
    atomic force microscopy (AFM), 198, 198f
    ichnograph, 199f–200f
    two-dimensional topological graph, 199–200, 199f–200f
  X-ray diffraction pattern, 99, 100f
  zeta (ζ) potential, 182f
    copper concentration, 183
    copper ions, 182–183, 184f
    isoelectric point (IEP), 182
    semiquantitative analysis, 184–185, 186t
    solid-liquid separation, 183
    X-ray photoelectron spectroscopy spectra, 184, 185f
Chromium oxide, 31
Cinnabar, 31
Component analysis methods, 52–53
  gas chromatography (GC), 55, 55f
  inductively coupled plasma mass spectrometry (ICP-MS), 53
  ion chromatography (IC), 53–54, 54f
  quadrupole mass spectrometry (QMS), 55–56, 56f

Component release
  chalcopyrite, 99–100, 114
    morphology, 107–113
    quartz and calcite, 114–118
    sphalerite, 118–123
  galena, 102, 133–134
  pyrite, 102–104, 134–148
  quartz, 100–101
  research methods
    high-resolution X-ray microfault three-dimensional imaging (HRXMT), 105–106
    inductively coupled plasma mass spectrometry (ICP-MS), 106
    IR-UV test, 105
  sphalerite, 100–101, 123–133
Copper sulfide, 98–99, 226
Crystal growth, 6, 20f
  chalcopyrite, 108
  discontinuity, 6
Crystalline melt inclusions, 23

D

Datuanshan Cu deposit, 83, 84t
Daughter minerals, 7, 10, 22, 59
  electron probe analysis, 80
  Fe-containing dolomite fluid inclusions, 78–80, 79t, 81t
  quartz veins, 2, 2f
  titanium-bearing dolomite fluid inclusions, 78, 78f
Density functional theory (DFT), 180
  chalcopyrite surface and copper
    adsorption structural parameters, 210–211, 210f
    $CuCl_2$ stable adsorption structure, 209–210, 210f
    difference charge density, 211, 211f
    Mulliken charge, 211
  crystal model and calculation method
    adsorption, 191
    chalcopyrite, 191–193
    interaction energy, 191
    Materials Studio (MS), 190
    pyrite, 194–195
    sphalerite, 193–194
    substitution effect, 191
    Vienna Ab Initio Simulation Package (VASP), 190
  pyrite surface and copper, 218–219, 218f–219f, 219t
  sphalerite surface and copper, 211–212
    adsorption energy and bandgap, 215, 215t
    average Mulliken population, 216, 217t
    Cu adsorption, 213, 214f
    Cu substitution, 212, 212f
    density of state of the atoms, 215–216, 216f
    electron densities, 216–217, 217f
    substitution energy, 213, 213t
  surface relaxation and reconstitution, 195
    chalcopyrite, 196–201
    pyrite, 206–209
    sphalerite, 201–206
Detection/composition study
  extraction and analysis, 48–66
  inclusions' salinity, 44–47
  modern research techniques, 39–44
  optical microscopy research, 29–39
Dongguashan Cu-Au deposit, 83, 86t–87t
Double-sided polished sheets
  general requirements, 29–30
  grinding process, 31
Duobuza Cu deposit, 70–74, 75t–76t

E

Electron probe microanalysis (EPMA), 78
Equilibrium theory calculation, of solubility, 160
  pH, 164t, 169f
    Debye-Hückel limit formula, 168
    minerals and compounds, 166t
    stability constant, 165t
  pure water, 160–161, 163t
    conditional solubility product, 161
    equilibrium constant, 161
    galena, 162
    total concentration, 162

F

Filling degree, 21
Fluid capture, 2
  changes, 8–14
  discontinuity, of crystal growth, 6
  inclusion formation, 6
  irregular crystal growth, 5, 5f
  solid-phase substances and impurities, 6–7
  temperature and pressure, 7–8
  unbalanced growth, 6

Fluid inclusion components
chemical composition
group inclusions, 50–56
individual inclusions, 57–66, 58t
issues, 49–50
Fluid melt inclusions, 23, 24f
Fourier transform infrared (FTIR), 62–63, 63f
Freezing method, 44
measuring instruments, 47
notes, 46–47
principle, 45–46

G

Galena
chemical composition, 102, 102t
component release of inclusions, 134
infrared optics and SEM/EDS, 133, 133f–135f
X-ray diffraction pattern, 102, 102f
zeta (ζ) potential, 188–189, 189f
Gangue, 98–99, 102
Gas chromatography (GC), 28–29, 55, 55f
Gas chromatography-mass spectrometry (GC-MS), 28–29
Gaseous-phase composition
Duobuza Cu deposit, 70–74, 75t–76t
Huayuan Pb-Zn deposit, 70, 71t–73t
Gas inclusions, 21
Generalized gradient approximation (GGA), 191
Genetic classification, 17, 18t
metamorphosed inclusions, 20
primary inclusions, 19–20, 19f–20f
pseudo-secondary inclusions, 20
secondary inclusions, 20
Grinding method, 51

H

Heavy metal elements, 80–83
Hellmann-Feynman force, 192
Heterogeneous fluid system, 7
High-resolution X-ray computed tomography (HRXCT), 44, 44f
High-resolution X-ray microfault three-dimensional imaging (HRXMT) analysis, 111, 111f–112f, 112t, 114t
Huayuan Pb-Zn deposits, 70, 71t–73t
Huize Pb-Zn mine, 100
Hydrocarbon fluid inclusions, 38–39
Hydrothermal fluid inclusion system, 8

I

Inclusion component fluids
cleaning, 51
individual mineral selection, 50–51
liquid-phase component, 52
opening inclusions, 51–52
Inclusion morphology detection methods, 28
Inclusion sheets
advantages and disadvantages, 29, 30t
double-sided polished sheets, 29–31
selection and production, 29
Inclusions' salinity
freezing method, 44
measuring instruments, 47
notes, 46–47
principle, 45–46
melting temperature, 44
microthermal method, 45
Individual fluid inclusions
destructive analysis
laser ablation inductively coupled plasma mass spectrometry (LA-ICP-MS), 57, 59f
scanning electron microscopy/energy-dispersive X-ray spectroscopy analysis (SEM/EDS), 59–60
secondary ion mass spectrometry analysis (SIMS), 60–61, 61f
nondestructive analysis
Fourier transform infrared (FTIR), 62–63, 63f
laser Raman spectroscopy (LRM), 61–62, 62f
microbeam proton-induced X/γ-ray analysis, 64–66, 65f
synchrotron radiation X-ray fluorescence (SXRF), 63–64
Inductively coupled plasma mass spectrometry (ICP-MS), 53, 106, 152
Infrared microscopy
forbidden band energy levels, 39–40, 41f
mineral fluid inclusions, 40–41
opaque mineral (pyrite), 39, 40f
working principle, 39–40
Ion chromatography (IC), 28–29, 53–54, 54f
Irregular distribution, 36
Irreversible volume change, 12
Isoelectric point (IEP), 182, 186–187

J

Jinding mine, 85–88, 89t–90t

L

Lanping/Jinding Pb-Zn deposit, 85–88, 89t–90t, 91f
Laser ablation inductively coupled plasma mass spectrometry (LA-ICP-MS), 57, 59f
Laser Raman spectroscopy (LRM), 61–62, 62f
Liquid inclusions, 21
Liquid-phase components, 69–70, 74–78, 75t–77t

M

Magmatic inclusions
  crystalline melt inclusions, 23
  fluid melt inclusions, 23, 24f
  vitreous melt inclusions, 23
Materials Studio (MS), 190
Mechanical crushing method, 51
Melt inclusions, 21
Metal components
  Cu-Au mineral, 83–85, 86t–87t
  ore-forming fluids, 80
    dilution multiples and concentration conversion, 82–83, 84t
    sample preparation, 82
    sample selection, 80
    sample test, 82
  Pb-Zn mineral, 85–88, 89t–90t, 91f
  quartz and pyrite, 88–95, 92t–95t
Metamorphosed inclusions, 20
Metastability, 11
Michelson interferometer, 62–63
Microbeam proton-induced X/γ-ray analysis, 64–66, 65f
Mineral crystals, 8
  growth and defects
    formation, 3–5
    line defects, 4
    particle stacking, 3–4, 4f
    point defects, 4
    volume defects, 4
Mineral surfaces, 172
  density functional theory (DFT), 190–219
  zeta (ζ) potential measurement, 180–189
Mineral fluid inclusions, 69
  definition, 1–3, 2f
  formation and mechanism
    fluid capture, 5–8
    mineral crystals, 3–5, 4f
  inevitability and universality, 15
  internal composition, 69–96
  phase changes, 8–11
  physical changes, 11–14
Molten silicate inclusions, 9–10, 14
Monkhorst-Pack, 193–195
Multielement analysis, 48

N

Near-infrared light, 28

O

Olivine, 13–14
Opaque-translucent mineral fluid inclusions, 28
Opening inclusions, 51–52
Optical microscopy research, 29
  identification, 31–33, 32f
  inclusion sheets
    advantages and disadvantages, 29, 30t
    double-sided polished sheets, 29–31
    selection and production, 29
  inclusions *vs.* impurities, 31, 32t
  morphology
    abundance, 36–37, 36f
    color, 37
    distribution, 35–36
    shape, 33
    size, 34–35
  phase state of matter
    gaseous phase, 38
    hydrocarbon, 38–39
    liquid phase, 38
    solid phase, 38
  positioning, 32, 32f–33f
Ordinary optical microscope, 28

P

Particle-induced X-ray emission (PIXE), 64–66
Perdew-Burke-Ernzerhof exchange (PBE), 192
Petroleum inclusions, 38–39
Phase boundary, 1–3
Phase changes, 8
  crystallization, body wall, 8–10, 9f
  daughter minerals, 10

immiscibility, 10–11
metastability, 11
shrinkage, 10–11
Physical changes, 11–12
fluid infiltration and loss, 13–14
shape, 12–13
volume change, 13
irreversible, 12
reversible, 12
Physical state classification, 17, 18t
fluid inclusions
carbon dioxide inclusions, 22
daughter mineral inclusions, 22
gas inclusions, 21
liquid inclusions, 21
oil-gas inclusions, 22, 22f
pure gas inclusions, 21
pure liquid inclusions, 21
magmatic inclusions, 23
Polarizing microscopy, 29
Polymetallic sulfide, 98
Primary inclusions, 19–20, 19f–20f
Projector augmented wave (PAW), 190, 192
Pseudo-secondary inclusions, 19f, 20
Pure gas inclusions, 21
Pure liquid inclusions, 21
Pyrite, 85, 88–95
component release, 141–148, 141t–142t, 143f, 144t, 147f, 148t
crystal model and calculation method, 194–195, 194f
Dapingzhang area of Yunnan, 104, 105t, 145–147, 145f–146f, 147t
high-resolution X-ray topography, 140, 140f–141f
infrared optical microscopic analysis, 134–136, 136f
scanning electron microscopy/energy-dispersive X-ray spectroscopy (SEM/EDS), 137–140, 137f, 139t
solubility
nonoxidative dissolution, 158–159, 158f–159f
oxidative dissolution, 158–159, 158f, 160f
surface relaxation and reconstitution
before and after relaxation, 206, 206f–207f
atomic coordination and displacements, 206, 206t, 208t
atomic force microscopy (AFM), 208, 208f
Weixin, Yunnan, 103–104, 103t, 104f
zeta (ζ) potential, 188–189, 189f
Pyroxene, 13–14

Q

Qinling pine groves, 23, 24f
Quadrupole mass spectrometry (QMS), 28–29, 55–56, 56f
Quartz, 1, 19–20, 19f, 37, 50–51, 88–95, 100–101, 114–118, 129–133
chemical composition, 100, 100t
Dapingzhang area of Yunnan, 104, 105t
Weixin, Yunnan, 103–104, 103t, 104f

R

Raman effect, 61
Raman spectral slope, 47
Raman spectrum, 61
Raoult's law, 45–46
Refrigerant-cooling freezer, 47
Regular distribution, 35
Reversible volume change, 12

S

Salinity. *See* Inclusions' salinity
Scanning electron microscopy (SEM), 41–43, 42f–43f
Scanning electron microscopy/energy-dispersive X-ray spectroscopy (SEM/EDS), 59–60, 78
chalcopyrite
inclusion domain, 109–110, 109f
residual position domains, 108–109, 108f
semiquantitative analysis, 109–110, 110t
galena, 133
pyrite, 137–140
sphalerite surface, 127–129
Schrodinger equation, 192
Secondary inclusions, 20
Secondary ion mass spectrometry analysis (SIMS), 60–61, 61f
Segmental ablation, 57
Solid-liquid separation, 183
Solid-phase composition, 78–80, 78f, 79t, 81t

Solubility
chalcopyrite
acidic conditions, 155
neutral and alkaline solutions, 155
vs. pH, 153, 154f
vs. time, 153, 154f
chemical equilibrium calculation, metal ions
Cu^{2+} taring and component distribution, 169–171, 171f
Pb^{2+} taring and component distribution, 173–176, 174f, 177f
Zn^{2+} taring and component distribution, 172–173
equilibrium theory calculation
pH, 164–169
pure water, 160–164
pyrite
nonoxidative dissolution, 158–159, 158f–159f
oxidative dissolution, 158–159, 158f, 160f
research methods, 152–153
sphalerite, 156–158, 156f–157f
Sphalerite, 31, 85, 98–99, 223–226, 224f, 225t
chalcopyrite
Cu, Fe, Cl^-, SO^{2-}_4, 121, 121t–122t
flotation separation methods, 118–119
fluid inclusion photomicrograph, 120, 120f
fluid inclusion sheets, 119, 119f
micrograph, 119, 119f–120f
chemical composition, 100, 101t
crystal model and calculation method, 193–194, 193f
freezing point and salinity value, 126–127, 127t
isomeride, 101
morphology and types
gas-rich phase inclusion, 124, 124f
group of microinclusions, 124, 124f–125f
high-resolution X-ray microfault 3D imaging, 124, 125f
liquid-gas inclusion, 124, 124f–125f
liquid-rich phase inclusion, 124, 124f
pure gaseous inclusion, 124, 124f–125f
pure liquid inclusion, 124, 124f–125f
thin sheets, 123, 123f
and quartz
dissolution, 131, 131t
Zn and Cl^-, 129–130, 130t, 132t
scanning electron microscopy/energy-dispersive X-ray spectroscopy (SEM/EDS), 127–129, 128f
solubility, 156–158, 156f–157f
surface relaxation and reconstitution, 202f
before and after geometry optimization, 201–202, 201f
atomic force microscopy (AFM), 204–205, 204f
geometric parameters, 202–204, 203t
structural optimization, 201–202
two-dimensional electron cloud topology, 204, 205f
zeta (ζ) potential
equilibrium potential, 187
isoelectric point (IEP), 186–187
vs. time, 187, 187f, 188t
Zn concentration, 187–188, 188t
Sulfide mineral flotation, 226–228
Supersaturated solution, 2
Surface adsorption
chalcopyrite, 182–186
pyrite and galena, 188–189
sphalerite, 186–188
Surface relaxation/reconstitution, sulfide mineral, 195
chalcopyrite, 196–201
pyrite, 206–209
sphalerite, 201–206
Synchrotron radiation X-ray fluorescence (SXRF), 63–64

T

Taibai Au deposit, 78–80, 79t, 81t
Thermal explosion method, 28–29, 51–52
Tianmashan S-Au deposit, 83, 84t
Time-of-flight-secondary ion mass spectrometry (TOF-SIMS), 60–61, 61f
Titanium-bearing dolomite fluid inclusions, 78, 78f
Topaz, 1
Transmission light electron microscopy, 29

U

Ultraviolet light microscopy, 39

V

Vacuum milling-gas chromatography, 70
Vienna Ab Initio Simulation Package (VASP), 190
Vitreous melt inclusions, 23

X

Xiaoqinling Au mine, 91
Xihuashan tungsten deposit, 74, 77t
X-ray diffraction (XRD) analysis, 99, 100f, 102

Z

Zeta (ζ) potential measurement, 180–181
  chalcopyrite, 182f
    copper concentration, 183
    copper ions, 182–183, 184f
    isoelectric point (IEP), 182
    semiquantitative analysis, 184–185, 186t
    solid-liquid separation, 183
    X-ray photoelectron spectroscopy spectra, 184, 185f
  pyrite and galena, 188–189, 189f
  sphalerite
    equilibrium potential, 187
    isoelectric point (IEP), 186–187
    vs. time, 187, 187f, 188t
    Zn concentration, 187–188, 188t
  test method, 181
ZetaProbe potentiometer, 152
ZnS supercell model, 194

FSC
www.fsc.org
MIX
Papier | Fördert gute Waldnutzung
FSC® C083411

Kolibri 360 GmbH
Ferdinand-Jühlke-Straße 7
99095 Erfurt, Deutschland
produktsicherheit@kolibri360.de